AF780954

International Basic Research Collaboration at the U.S. Department of Defense

An Overview

ALISON K. HOTTES, MARJORY S. BLUMENTHAL, JARED MONDSCHEIN,
MATTHEW SARGENT, CAROLINE WESSON

Prepared for the Office of the Secretary of Defense
Approved for public release; distribution unlimited

RAND NATIONAL DEFENSE RESEARCH INSTITUTE

For more information on this publication, visit **www.rand.org/t/RRA1579-1**.

About RAND

The RAND Corporation is a research organization that develops solutions to public policy challenges to help make communities throughout the world safer and more secure, healthier and more prosperous. RAND is nonprofit, nonpartisan, and committed to the public interest. To learn more about RAND, visit www.rand.org.

Research Integrity

Our mission to help improve policy and decisionmaking through research and analysis is enabled through our core values of quality and objectivity and our unwavering commitment to the highest level of integrity and ethical behavior. To help ensure our research and analysis are rigorous, objective, and nonpartisan, we subject our research publications to a robust and exacting quality-assurance process; avoid both the appearance and reality of financial and other conflicts of interest through staff training, project screening, and a policy of mandatory disclosure; and pursue transparency in our research engagements through our commitment to the open publication of our research findings and recommendations, disclosure of the source of funding of published research, and policies to ensure intellectual independence. For more information, visit www.rand.org/about/principles.

RAND's publications do not necessarily reflect the opinions of its research clients and sponsors.

Published by the RAND Corporation, Santa Monica, Calif.
© 2023 RAND Corporation
RAND® is a registered trademark.

Library of Congress Cataloging-in-Publication Data is available for this publication.

ISBN: 978-1-9774-1059-7

Cover photo: John Williams/ Office of Naval Research.

Limited Print and Electronic Distribution Rights

This publication and trademark(s) contained herein are protected by law. This representation of RAND intellectual property is provided for noncommercial use only. Unauthorized posting of this publication online is prohibited; linking directly to its webpage on rand.org is encouraged. Permission is required from RAND to reproduce, or reuse in another form, any of its research products for commercial purposes. For information on reprint and reuse permissions, please visit www.rand.org/pubs/permissions.

About This Report

The U.S. Department of Defense (DoD) invests in basic research to encourage progress in science and technology fields that are likely to contribute to national defense. DoD also has a long history of international engagement on basic research. While the United States, including DoD, has traditionally supported openness in basic research, concerns are growing that U.S. openness in basic research might be unduly aiding strategic competitors, particularly China. Recognizing the importance of basic research to its mission and the changing risks and benefits that it faces, DoD seeks a better understanding of its place in the international basic research ecosystem and strategic and scientific considerations for DoD international basic research collaborations. DoD is also interested in suggestions for improving how it uses international basic research collaborations.

The research reported here was completed in January 2022 and underwent security review with the sponsor and the Defense Office of Prepublication and Security Review before public release.

RAND National Security Research Division

This research was sponsored by the Basic Research Office within the Under Secretary of Defense for Research and Engineering and conducted within the Acquisition and Technology Policy Program of the RAND National Security Research Division (NSRD), which operates the National Defense Research Institute (NDRI), a federally funded research and development center sponsored by the Office of the Secretary of Defense, the Joint Staff, the Unified Combatant Commands, the Navy, the Marine Corps, the defense agencies, and the defense intelligence enterprise.

For more information on the RAND Acquisition and Technology Policy Program, see www.rand.org/nsrd/atp or contact the director (contact information is provided on the webpage).

Acknowledgments

We thank our sponsor, Bindu Nair, and her staff for guidance in formulating the project and their assistance in identifying prospective interviewees. We are grateful for the research support provided by Kelly Eusebi, Daniella Henry, and Sale Lilly and appreciate the assistance preparing the manuscript that Nora Spiering, Sydne Newberry, Emily Payne, Evan Smith, and Stacey Smith provided. We thank the individuals throughout the Department of Defense, the federal government, and academia who participated in interviews. We also thank Erin Fitzgerald, Richard Silberglitt, and John Parmentola for reviewing a draft of the report and Michael Kennedy, Christopher Mouton, Chad Ohlandt, Steven Popper, and Jim Powers for providing constructive comments during in-progress reviews.

Executive Summary

The U.S. Department of Defense (DoD) defines basic research as "systematic study directed toward greater knowledge or understanding of the fundamental aspects of phenomena and of observable facts without specific applications toward processes or products in mind."[1] Despite the lack of focus on application, basic research provides the essential foundation for *all* technical innovation.

The research enterprise is becoming increasingly global, with the share of the world's research and development that the United States funds decreasing from 69 percent in 1960 to 28 percent in 2016.[2] The overall increase in basic research investment worldwide is also reducing the share of research—and basic research—that the United States funds.[3] The growth of the worldwide research enterprise presents both opportunities and risks to the United States and DoD. On the positive side, international basic research collaboration (IBRC) can enable cost-sharing with partners and allies; provide access to unique, world-class infrastructure, resources, and expertise; and build network connections that provide visibility into cutting-edge advances and reduce the potential for technological surprise. IBRC can also help the United States promulgate its approach to research ethics, norms, and values and can serve as a starting point for engaging with other countries and building relationships that can provide opportunities to influence areas beyond basic research. On the negative side, IBRC can position partners to compete with the United States to turn the resulting discoveries into defensive and economic advantages. However, excessive restrictions on scientific international engagement could, paradoxically, reduce American competitiveness in critical technologies and weaken the U.S. defense industrial base.[4] Multiple studies have concluded that keeping results of fundamental (basic and non-proprietary applied research) research largely unrestricted is in the best interests of the United States, even when the actions of adversaries are considered.[5]

However, concerns are growing that the research strength and practices of strategic competitors, particularly the People's Republic of China (PRC), threaten the position of the United States as a partner of choice for some foreign researchers. In response, the Basic Research

---

[1] Code of Federal Regulations, Title 32, Section 272.3.

[2] John F. Sargent, Jr., *Global Research and Development Expenditures: Fact Sheet*, Congressional Research Service, June 27, 2018. The percentages refer to all research and development, not just basic research.

[3] Defense Science Board, *Report of the Defense Science Board Task Force on Basic Research*, Office of the Under Secretary for Defense for Acquisition, Technology, and Logistics, January 2012.

[4] See National Research Council, *Beyond "Fortress America": National Security Controls on Science and Technology in a Globalized World*, National Academies Press, 2009.

[5] See, for example, National Academy of Engineering, *Scientific Communication and National Security*, National Academies Press, 1982, pp. 3-4; and JASON, *Fundamental Research Security*, MITRE Corporation, December 2019.

Office (BRO) within the Office of the Under Secretary of Defense for Research and Engineering asked the RAND National Defense Research Institute to

1. summarize how DoD approaches IBRC and how DoD fits into the broader research ecosystem[6]
2. identify scientific and strategic considerations for IBRC for DoD
3. formulate suggestions to help DoD improve how it uses IBRC.

The remainder of this summary describes the findings for the three tasks above.

Overview of Department of Defense Approaches to International Basic Research Collaboration

In fiscal year (FY) 2018, DoD obligated $2.4 billion to basic research, which was 4.5 percent of DoD's research and development (R&D) obligations.[7] DoD's main basic research performers are academic institutions (54 percent) and its own intramural laboratories (22 percent).[8] Foreign entities receive 3 percent of DoD's basic research obligations, and federally funded research and development centers (FFRDCs), other nonprofits, and industry receive the remaining 21 percent.[9]

The Army, the Navy, and the Air Force all station personnel around the world who are responsible for understanding the state of research on multiple scientific topics in particular geographic regions. These personnel have a variety of titles (e.g., science director, international program manager). For simplicity, this report refers to the group collectively as *international office persons* (IOPs). IOPs merge their knowledge of the international research landscape with their understanding of their services' current and future R&D activities and needs across the spectrum of technology readiness levels to identify international expertise of potential benefit to their services. DoD IOPs are the face of DoD basic research to many around the world and were thus a focus of the project.

DoD IOPs conduct multiple types of exploratory activities to understand the international research ecosystem and to identify funding opportunities, including reading the literature,

---

[6] The majority of the analysis considered all of DoD, including the Army, the Navy, and the Air Force, collectively. The analysis did not examine the specific roles, responsibilities, and authorities of entities within DoD, such as BRO.

[7] National Center for Science and Engineering Statistics, *Survey of Federal Funds for Research and Development, Fiscal Years 2018–19*, January 2020, Table 5. The value includes obligations from the defense agencies and the services. FY 2018 was the most recent year available at the time of the analysis.

[8] National Center for Science and Engineering Statistics, *Survey of Federal Funds for Research and Development, Fiscal Years 2018–19*, January 2020, Table 30.

[9] National Center for Science and Engineering Statistics, *Survey of Federal Funds for Research and Development, Fiscal Years 2018–19*, January 2020, Table 30.

attending conferences, and arranging (prior to the coronavirus pandemic) liaison visits to laboratories of foreign principal investigators (PIs). DoD IOPs also fund a range of activities, including grants to international PIs, travel for international PIs to meet with service personnel, and international workshops and conferences. While DoD IOPs spend the bulk of their time engaging with the academic community, they also engage with officials from international organizations and partner government civilian and defense organizations to share information about basic research investments and to support a range of engagements.

Beyond the work of IOPs, DoD facilitates and enables IBRC in multiple ways.

- DoD and foreign government partners (not necessarily ministries of defense) issue joint calls for research projects that include team members from each country.[10]
- DoD entities run several international exchange programs for researchers.
- Many intramural DoD researchers collaborate with international colleagues.
- Some programs allow foreign universities to submit applications.
- The services have several programs that, although not focused on foreign researchers, can be used by visiting foreign researchers to access DoD intramural laboratories.

Comparisons Between How the Department of Defense and Other Organizations Approach International Basic Research Collaboration

The project team used open sources and interviews to compare DoD's approach to IBRC with the approaches of four example government research funding organizations: the U.S. Department of Energy (DOE), the U.S. National Science Foundation (NSF), the Agence Nationale de la Recherche (National Research Agency of France) (ANR), and the National Natural Science Foundation of China (NSFC). The analysis helped the project team understand the range of IBRC approaches in use and the location of DoD in the basic research ecosystem:

1. **Among the organizations examined, DoD is unique in its emphasis on the top-down approach of directly funding foreign researchers in foreign institutions in the absence of a project with domestic researchers.** The approach is part of a long-term strategy to develop relationships with promising researchers, usually early in their careers, and ultimately integrate them into the DoD research enterprise, which is consistent with DoD's mission focus.[11] DoD's ability to directly fund foreign researchers allows DoD to act unilaterally, providing flexibility and agility.[12]

---

[10] In joint calls, each research team typically submits only a single proposal, and each sponsor funds the researchers from its country.

[11] Other approaches could also be appropriate for a mission-focused organization.

[12] Bottom-up approaches, such as the uncoordinated funding of joint projects, where teams self-organize and each partner finds funding independently, can also provide flexibility and agility. This project did not assess the possible utility of organizations changing the emphasis that they put on specific funding mechanisms.

2. **DoD grants permit publication, offer favorable intellectual property terms, and offer the prestige of collaborating with U.S. scientists, which makes DoD the partner of choice for many foreign researchers.** In contrast, China often offers more generous amounts of funding but attaches far more substantial constraints and conditions to the funds it provides, such as transferring intellectual property from U.S.-funded work to China and hiding the affiliation with China.[13]
3. **DoD's approach to engaging with non-U.S. researchers heavily leverages the worldwide presence of IOPs, who engage with promising foreign researchers to increase the relevance of their grant applications to DoD and acquire direct knowledge about international research capabilities and opportunities.**[14] While DOE also stations some employees abroad, they support all DOE activities, not just research. In contrast, NSF closed its overseas offices in 2018, indicating that changes in the world had reduced the need to have a physical presence in a location.[15] NSF instead sends teams of experts abroad for short visits.[16]

Challenges to International Basic Research Collaboration Success at the U.S. Department of Defense

The project team identified three challenges to IBRC success at DoD.

1. **IOPs, who work primarily top-down, lack efficient and effective ways to assemble the information needed to identify and evaluate collaboration opportunities.**[17] Challenges that interviewees cited include lack of integration among DoD databases of grants and international agreements and other resources; lack of a central repository for documentation, such as trip reports from IOPs and researchers; reliance on personal connections for information-sharing; and loss of knowledge when IOPs rotate into new positions. The increase in basic research and quality basic research outside of the United States increases the amount of basic research activity that the DoD research enterprise needs to track. Thus, efficient processes that can be scaled to allow DoD IOPs to keep up with the growing amount of quality international research will likely become increasingly valuable going forward.

---

[13] U.S. Senate Permanent Subcommittee on Investigations, "Threats to the U.S. Research Enterprise: China's Talent Recruitment Plans," staff report, 2019.

[14] This project did not assess the possible utility of organizations changing the ways they interact with researchers outside of the United States.

[15] NSF, "NSF Announces Changes to Overseas Offices," press release 18-003, February 21, 2018.

[16] NSF, Office of International Science and Engineering (OISE), "NSF MULTIPLIER—MULTIPlying Impact Leveraging International Expertise in Research Missions," webpage, undated.

[17] This project did not analyze the extent to which the lack of integrated data sources impacts intramural and extramural DoD researchers, who identify opportunity primarily in a bottom-up manner.

2. **Some foreign researchers choose not to work for DoD for a variety of personal, institutional, and country-wide reasons.** DoD's connection to defense deters some researchers. The administrative effort needed to obtain a DoD grant—particularly in view of the grants' small size—discourages other foreign researchers from applying.
3. **Some DoD intramural researchers prefer to avoid engaging internationally.** This reluctance has multiple sources, including concerns about unintentionally violating rules or causing a security breach and the time and effort needed to initiate and conduct IBRCs. Collaboration-related overhead, including the time needed to navigate cultural and language barriers and obtain approvals and funding through multiple bureaucracies, can discourage both DoD and non-DoD researchers from IBRC and offset the benefits of collaborating, particularly in fast-moving fields. (The concerns, particularly the first one, are also relevant to academic researchers, regardless of whether they use DoD funds.[18])

Scientific and Strategic Considerations for IBRC

Interviewees and the literature identified multiple factors that influence the value of IBRC opportunities (Table S.1).[19] *Scientific considerations* predominantly relate to advancing science and to the benefits a collaboration might provide. *Strategic considerations* are primarily nonscientific and include factors such as leadership desires to strengthen ties with another country or to avoid strengthening a competitor's military capabilities. While some strategic considerations can be connected to benefits, such as supporting economic growth, strategic considerations more commonly characterize the risks of a collaboration. For example, working with a potentially untrustworthy partner in a scientific field where breakthroughs could rapidly be applied to create a military advantage is risky. Table S.1 provides additional examples of benefits and risks. Each collaboration is unique, and different considerations can dominate decisionmaking processes about whether to proceed with individual opportunities.

Suggestions to Help the U.S. Department of Defense Improve How It Uses International Basic Research Collaboration

The project team offers five suggestions that might help DoD improve how it uses IBRC.

18 The project did not compare attitudes towards international engagement among DoD intramural researchers, DoD-funded grantees, or U.S. basic researchers supported by other parts of the U.S. government.

19 Another factor, which is relevant for all collaborations, is the extent to which a collaboration would fit into a researcher's or an organization's overall research needs.

TABLE S.1
Considerations for International Basic Research Collaboration

Consideration	Examples	Scientific or Strategic?
Access to unique and diverse resources	• Obtaining access to exceptional and distinctive expertise, facilities, tools, and materials (benefit)	Scientific
Ability to combine resources and share the burden	• Addressing global challenges collectively (benefit) • Building and maintaining large-scale scientific instruments and infrastructure (benefit) • Pooling project resources (benefit) • Leveraging other countries' prior investments (benefit) • Synchronizing research priorities (benefit)	Both
Build and strengthen relationships	• Forming researcher-to-researcher, institution-to-institution, and government-to-government relationships that can - provide awareness to reduce technological surprise and suggest transition opportunities (benefit) - create influence to shape scientific norms (benefit) - facilitate science diplomacy (benefit)	Both
Trust in a partner[a]	• Assessment of partner's intent and capability to use research results in ways that are consistent with the wishes of other partners (risk)	Strategic
Economic impacts	• Ability to familiarize a government with another country's innovation landscape and the associated opportunities for future exports (benefit) • Support of long-term economic growth (benefit) • Potential to spread benefits from resulting commercial applications among multiple countries (risk, but possibly a benefit with select partners)	Strategic
Defense impacts	• Ability of a scientific breakthrough to support applications that provide a military advantage (risk, but possibly a benefit with select partners) • Potential to support the formation of supply chains in competitors and weaken the domestic defense industrial base (risk)	Strategic
Maturity of a scientific field	• Potential for particularly useful visibility stemming from partnering in immature fields to explore multiple lines of inquiry (benefit)	Both
Speed of transition within a scientific field	• Potential for rapid application to exacerbate defense and economic impacts (risk, but possibly a benefit with select partners) • Potential for rapid transition of a project from basic to applied research to trigger export restrictions unexpectedly (risk)	Strategic
Amount of collaboration-related overhead	• Regulations and bureaucracies of each partner (risk) • Effort needed to establish the terms of a collaboration (risk) • Language and culture differences (risk)	Strategic

[a] Trust matters at the individual, organizational, and country levels.

1. **Provide knowledge management tools to help DoD research funders and DoD intramural researchers identify, evaluate, establish, and conduct international collaborations.** Such tools would ideally make a variety of resources easily discoverable and usable, such as IOP site visit reports, international trip reports from DoD intramural researchers, results from service studies that analyze the international landscape, and collaboration guidance. Tools could also provide visibility into contacts and engagements and allow users to leverage data from across DoD (and ideally the U.S. government).
2. **Review, improve, and publicize policies and guidance to reduce instances of DoD intramural and DoD-funded U.S. academic researchers unnecessarily choosing not to work with high-quality foreign partners.** Particularly for U.S. academics whose research is supported by multiple services, DoD-wide guidance would be especially useful. As of October 2021, DoD was updating its policies and processes for international research collaboration.[20]
3. **Streamline application and approval processes to reduce overhead associated with IBRC.** The amount of effort needed for bureaucratic processes can deter both international researchers from applying for DoD grants and DoD intramural researchers from seeking international partners.
4. **Analyze whether adjusting the size, number, and terms of DoD grants to foreign researchers could provide a competitive advantage and help DoD meet its goal of being the partner of choice.** The analysis might also consider the overall amount of basic research funds that DoD should invest in supporting international researchers in view of competing priorities.
5. **Assemble a fuller, multiservice picture of DoD international basic research efforts and processes; in view of the growing quantity and quality of research being conducted outside of the United States and the increased technological competition with the PRC and others, such information could help DoD assess and update its IBRC processes, policies, and strategies.** DoD might gather additional information (ideally in a way that minimizes the associated overhead) about the amount that DoD obligates to the U.S. side of IBRCs, how DoD intramural and DoD-funded U.S. academic basic researchers identify and evaluate international opportunities, and challenges that DoD intramural and extramural researchers are encountering. Other topics of interest might include the relative effectiveness of the processes that DoD researchers and IOPs use to identify and evaluate IBRC opportunities, the extent to which their efforts complement each other and collectively allow DoD sufficient visibility into—and access to—the basic research ecosystem, the actual and desired relationship between DoD's international and domestic basic research portfolios, and opportunities for improvement.

[20] BRO staff person, email to A. Hottes, October 29, 2021.

Contents

Figures and Tables

Figures

Tables

CHAPTER ONE

Introduction

Basic research is the "systematic study to gain knowledge or understanding of the fundamental aspects of phenomena and of observable facts without specific applications toward processes or products in mind."[1] Basic research is a small portion of research and development (R&D), with only 16.4 percent (108 billion U.S. dollars) of the United States' 2019 total research expenditures of 658 billion U.S. dollars going to basic research.[2] Nevertheless, basic research provides a foundation for innovation, and, as the National Science Board (NSB) explained in 2008:

> The scientific and technological advances that have led to our Nation's remarkable ability to create new industries and jobs, improve the standard of living for people, and provide sophisticated technology that ensures our national security can be traced back to the outcomes of basic research.[3]

While basic research inherently supports multiple applications, U.S. Department of Defense (DoD) personnel often select projects with an awareness of their potential near- and long-term defense applications.[4]

The basic research enterprise is decentralized, with many researchers entrepreneurially seeking support from a range of government and private organizations. Today, many of the people who conduct basic research in the United States work in universities, and the higher education sector performed 45.7 percent of basic research in the United States in 2019.[5] In

---

[1] Federal Accounting Standards Advisory Board, *FASAB Handbook of Federal Accounting Standards and Other Pronouncements, as Amended*, Version 19, June 30, 2020, Appendix E, p. 9.

[2] Organisation for Economic Co-operation and Development (OECD), "Gross Domestic Expenditure on R&D by Sector of Performance and Type of R&D," *OECD.Stat*, undated.

[3] NSB, National Science Foundation, *Research and Development: Essential Foundation for U.S. Competitiveness in a Global Economy—A Companion to Science and Engineering Indicators—2008*, NSB-08-3, 2008, p. 1.

[4] Basic research is one of three types of science and technology (S&T) work that DoD supports; applied research and advanced technology development are the other two.

[5] OECD, "Gross Domestic Expenditure on R&D by Sector of Performance and Type of R&D," *OECD.Stat*, undated.

addition, some work in corporate or nonprofit private laboratories,[6] typically with private funding but sometimes also with government contracts or grants. The business enterprise and private nonprofit sectors performed 29.7 and 13.4 percent, respectively, of U.S. basic research in 2019.[7] Others who do basic research work in laboratories associated with government agencies such as DoD, the Department of Energy (DOE), and the U.S. Department of Health and Human Services (HHS). The government sector performed 11.2 percent of U.S. basic research in 2019.[8] The success of U.S. R&D has been attributed to what some have called the *triple helix*, the flow of people and funding over time among government, universities, and industry as an idea evolves from its earliest expression in basic research to its ultimate application.[9] Preparations to transition discoveries from DoD-funded basic research to the defense research enterprise often start early with program managers and others bringing work to the attention of potential transition partners. In contrast with the openness of basic research, some basic research collaborators—particularly non-U.S. collaborators—are not eligible to participate in some transitions.

1.1. Types of International Research Collaboration

The research enterprise is becoming increasingly global, with the share of the world's R&D that the United States funds decreasing from 69 percent in 1960 to 28 percent in 2016.[10] The overall increase in basic research investment worldwide is also reducing the share of research—and basic research—that the United States funds.[11] Because more countries report on their total research expenditures than report on their basic research expenditures, comparable figures for basic research are not available.[12] Figure 1.1, however, illustrates the phenomenon. U.S. expenditures on basic research in constant 2015 U.S. dollars increased 26 percent from 2011 to 2019, while the basic research expenditures of the People's Republic of China (PRC) increased

6 Examples of nonprofit private laboratories include the Allen Institute, the Allen Institute for AI, and the J. Craig Venter Institute.

7 OECD, "Gross Domestic Expenditure on R&D by Sector of Performance and Type of R&D," *OECD.Stat*, undated.

8 OECD, "Gross Domestic Expenditure on R&D by Sector of Performance and Type of R&D," *OECD.Stat*, undated.

9 National Research Council, *Funding a Revolution: Government Support for Computing Research*, National Academies Press, 1999.

10 John F. Sargent, Jr., *Global Research and Development Expenditures: Fact Sheet*, Congressional Research Service, June 27, 2018. The percentages refer to all R&D, not just basic research.

11 Defense Science Board (DSB), *Report of the Defense Science Board Task Force on Basic Research*, Office of the Under Secretary for Defense for Acquisition, Technology, and Logistics, January 2012.

12 For example, *OECD.Stat* reports total research expenditure for 44 countries for at least one year between 2011 and 2020 but the basic research expenditures for only 38 countries during that time. OECD, "Gross Domestic Expenditure on R&D by Sector of Performance and Type of R&D," *OECD.Stat*, undated.

FIGURE 1.1
U.S. and PRC Research Expenditures, 2011–2019

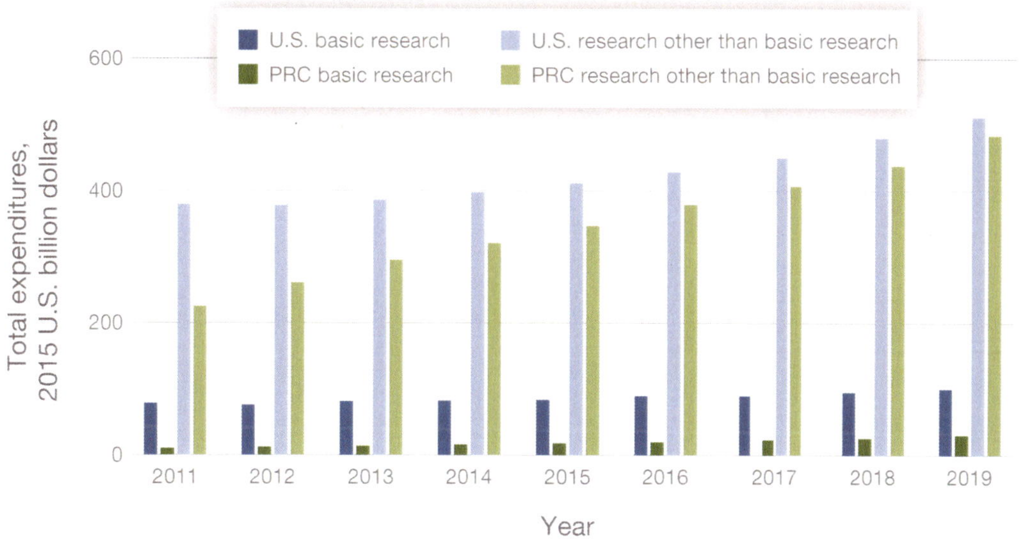

SOURCE: OECD, "Gross Domestic Expenditure on R&D by Sector of Performance and Type of R&D," *OECD.Stat*, undated.

176 percent.[13] In absolute terms, however, the PRC's 2019 basic research expenditures remained just a fraction of the United States' (31 billion U.S. dollars versus 100 billion U.S. dollars), while the PRC's 2019 expenditures for research other than basic research approached those of the United States (484 billion U.S. dollars versus 512 U.S. dollars).[14]

From DoD's perspective, good science has long been done in other countries. Acknowledgment of European research strength following World War II was a motivation for DoD to begin engaging with researchers in Europe.[15]

The growth of the worldwide research enterprise presents both opportunities and risks to the United States and DoD. On the positive side, international basic research collaboration (IBRC) can enable cost-sharing with partners and allies; provide access to unique infrastructure, resources, and expertise; and build network connections that provide visibility into cutting-edge advances and reduce the potential for technological surprise. On the negative side, IBRC can position partners to turn the resulting discoveries into defensive and economic advantages (provided they have sufficient absorption capacities). However, excessive

[13] OECD, "Gross Domestic Expenditure on R&D by Sector of Performance and Type of R&D," *OECD.Stat*, undated.

[14] OECD, "Gross Domestic Expenditure on R&D by Sector of Performance and Type of R&D," *OECD.Stat*, undated.

[15] Robert F. Phillips, *Administrative History of the European Office of Aerospace Research and Development, 1952–1975*, Air Force Systems Command, 1977.

restrictions on scientific international engagement could, paradoxically, reduce American competitiveness in critical technologies and weaken the U.S. defense industrial base.[16]

International research cooperation activities can take multiple forms. For example, Wagner, 2006, presents a framework that organizes activities along two dimensions: (1) the degree to which the physical work is distributed or centralized and (2) the degree to which the activity emerged spontaneously though the actions of researchers (bottom-up) or was organized by institutions or governments in a top-down process.[17] A project where the work occurs in multiple locations and results are shared virtually is highly decentralized; a project where the team works at a single facility is highly centralized. The top-down versus bottom-up dimension is sometimes also characterized as formal versus informal.[18]

Both dimensions are spectrums, and activities frequently have both centralized and distributed elements and top-down and bottom-up aspects. For example, the European Organization for Nuclear Research (CERN) gathers researchers in a single place, but data analysis can occur at multiple locations. Similarly, researchers might meet at a conference that institutions organized in a top-down process and decide to share data in a bottom-up manner.

The bilateral S&T cooperation agreements that the U.S. government (USG) has with other countries illustrate the variety of research cooperation activities, including

- coordinated or joint research projects, task forces, and studies
- joint organization of scientific seminars, conferences, symposia, and workshops
- training of scientists and technical experts
- exchanges or sharing of equipment and materials
- use of facilities and equipment
- development of technical standards
- visits and exchanges of scientists, engineers, or other appropriate personnel
- exchanges of scientific and technological information, as well as information on practices, laws, standards, and programs relevant to cooperation
- educational exchanges related to science, technology, and engineering
- establishment of science and innovation–based public-private partnerships.[19]

Table 1.1 summarizes ways government organizations can support and facilitate IBRC.

[16] See National Research Council, *Beyond "Fortress America": National Security Controls on Science and Technology in a Globalized World*, National Academies Press, 2009.

[17] Caroline S. Wagner, "International Collaboration in Science and Technology: Promises and Pitfalls," in Louk Box and Rutger Engelhard, eds., *Science and Technology Policy for Development, Dialogues at the Interface*, Anthem Press, 2006.

[18] Luke Georghiou, "Global Cooperation in Research," *Research Policy*, Vol. 27, No. 6, September 1998.

[19] The team compiled the list from a sample of U.S. bilateral agreements on scientific cooperation. The team chose the set to include geographic diversity and countries at varying levels of development. The set included agreements between the United States and the following countries: Germany (both overarching agreement and agreement on civil/homeland security matters), France, India, Kazakhstan, Romania, Australia, and the United Kingdom.

TABLE 1.1
Ways Government Organizations Support and Facilitate IBRC

Category	Description	Examples
Sponsor research or exploratory activities with an international component	Activities with a mandatory international connection that was deliberately and intentionally incorporated	• Issue calls for research proposals where domestic applicants must have an international partner (even if the partner is funded by its own government) • Contribute to international mega-science projects • Host conferences domestically or abroad where international researchers will be present • Send personnel abroad (e.g., through short visits or staffing overseas offices) to conduct technology scouting and understand the foreign research landscape • Create researcher exchange programs with foreign governments • Issue calls for research proposals that are only open to foreign researchers
Permit and enable the national research enterprise to engage internationally	Activities that allow but do not require international engagement by opening domestic research personnel, facilities, funding, and other resources to international researchers	• Issue calls for research proposals that allow domestic researchers to have an international partner • Permit grantees to make sub-awards to foreign entities • Allow foreign researchers to use domestic facilities • Allow domestic researchers to conduct work abroad • Allow researchers to present at international conferences • Permit foreign researchers to enter the country to attend domestic conferences • Allow foreign researchers to respond to calls for proposals, rather than restricting calls to domestic researchers
Work with partner governments and international organizations to facilitate IBRC	Activities that facilitate those in the other categories	• Develop formal, overarching agreements that provide ground rules (e.g., policies for intellectual property [IP], facility use, etc.) for collaborations, which can lower start-up costs for researchers and institutions • Develop joint funding calls or S&T roadmaps • Discuss ways to harmonize processes to facilitate collaboration

1.2. Role of Research Integrity in International Basic Research Collaboration

Knowledge acquired from basic research is a public good. It is nonrival (one person's use does not affect another's) and, absent unusual measures such as national security classification, nonexcludable (anyone can use it). Thus, scientists and science-policy analysts frequently argue that basic research competition should be viewed as a positive-sum rather than a zero-sum game.[20]

The positive-sum framework and the associated mutual advantage resulting from IBRC are tightly coupled to core research values. In 2019, then-director of the U.S. Office of Science and Technology Policy (OSTP) Kelvin Droegemeier referred to "free and open inquiry,

[20] Jenny J. Lee and John P. Haupt, "Winners and Losers in US-China Scientific Research Collaboration," *Higher Education*, Vol. 80, July 6, 2020.

reciprocity with other countries, people coming to America and using our laboratories working with us hand-in-hand" as U.S. values that complement the research values of integrity, honesty, transparency, openness, and mutual respect.[21] He observed that the values of the research community and of nation-states should, but do not always, align—even in the United States.[22] Historic research leadership has helped the United States to propagate its approach to ethics, norms, and values, inculcating these both in U.S. researchers and those who train or collaborate with them.

The OSTP has worked to increase awareness of how violations to core research principles, such as the suspected cybertheft of 31.5 terabytes of academic data by Iranian government–sponsored hackers and unreported participation in foreign government–sponsored talent recruitment programs, threaten the research enterprise and national and economic security.[23] According to the Federal Bureau of Investigation, China runs hundreds of talent recruitment programs "designed to acquire foreign technologies needed to augment China's comprehensive national strategy."[24] While some are legal, others offer incentives for "undisclosed and illicit transfers of technology."[25] The OSTP has been particularly concerned about participation in foreign government talent recruitment programs that create conflicts of commitment and interest and hidden diversions of results and resources and has argued that only with transparency and full disclosure is it possible to properly assess risks.[26] Other concerns include researchers who might not be who they appear to be, such as students sent to the United States by the Chinese military who do not disclose their military affiliation.[27] DoD has emphasized that disclosure alone is insufficient and that additional policies are needed to define what is acceptable and to reduce the risk of misappropriation.[28]

---

21 Joint Committee on the Research Environment, National Science and Technology Council, Executive Office of the President, *Summary of the 2019 White House Summit of the Joint Committee on the Research Environment (JCORE)*, November 2019, p. 2.

22 Joint Committee on the Research Environment, National Science and Technology Council, Executive Office of the President, *Summary of the 2019 White House Summit of the Joint Committee on the Research Environment (JCORE)*, November 2019.

23 OSTP, "Enhancing the Security and Integrity of America's Research Enterprise," briefing, July 2020.

24 Federal Bureau of Investigation, "Foreign Government-Sponsored Talent Recruitment Plans, Such as China's Talent Plans, Incentivize Economic Espionage and Theft of Trade Secrets," PSA Number 20200716-2, July 16, 2020.

25 Federal Bureau of Investigation, "Foreign Government-Sponsored Talent Recruitment Plans, Such as China's Talent Plans, Incentivize Economic Espionage and Theft of Trade Secrets," PSA Number 20200716-2, July 16, 2020.

26 OSTP, "Enhancing the Security and Integrity of America's Research Enterprise," briefing, July 2020.

27 Kate O'Keeffe and Melissa Korn, "China's Military Sends More Scholars Abroad, at Times Without Schools' Knowledge," *Wall Street Journal*, October 29, 2018.

28 U.S. Government Accountability Office, *Federal Research: Agencies Need to Enhance Policies to Address Foreign Influence*, GAO-21-130, December 2020, p. 39.

China's increasing research strength and use of practices at odds with traditional U.S. research values have motivated academic institutions[29] and the USG to increase their focus on research integrity and to make explicit some principles for research integrity for individual researchers and governments that formerly were often taken for granted.[30] For example, DoD expanded the disclosures required in response to its notices of funding opportunities, illustrating growing funder interest in developing a more complete picture of how key research personnel are supported.[31] The National Science Foundation (NSF) and DOE issued or clarified policies that preclude their own personnel from participating in foreign talent-recruitment programs while emphasizing values such as openness.[32] In January 2021, President Donald Trump issued National Security Presidential Memorandum (NSPM)-33, which prohibits federal employees from participating in talent recruitment programs sponsored by foreign governments and directs federal organizations to develop standardized processes and forms for research-related disclosures.[33] The Council on Governmental Relations identified a set of risks to consider when evaluating possible international research collaborations.[34] Similarly, the Association of American Universities, also focusing on enhanced disclosure, acknowledged concerns such as targeted cyberattacks, theft of intellectual capital, and "other forms of undue foreign government interference relating to research performed at American universities."[35] DoD has applauded such collective action.[36]

Multiple studies separated by decades have concluded that, even when the actions of adversaries are considered, keeping results of fundamental research (basic and nonproprietary applied research) largely unrestricted is in the best interests of the United States. In

---

[29] Major federal research funders "rely on universities to monitor financial conflicts of interest and collect information that could be used to determine the existence of non-financial conflicts" (U.S. Government Accountability Office, *Federal Research: Agencies Need to Enhance Policies to Address Foreign Influence*, GAO-21-130, December 2020, p. 16).

[30] Ethics for basic research can vary by country and discipline, although there are common principles. See Cortney Weinbaum, Eric Landree, Marjory S. Blumenthal, Tepring Piquado, and Carlos Ignacio Gutierrez Gaviria, *Ethics in Scientific Research: An Examination of Ethical Principles and Emerging Topics*, RAND Corporation, RR-2912-IARPA, 2019.

[31] Michael D. Griffin, *Actions for the Protection of Intellectual Property, Controlled Information, Key Personnel and Critical Technologies,* Memorandum for Under Secretary of Defense for Acquisition and Sustainment, Assistant Secretary of Defense for Acquisition, Service Acquisition Executives, Special Operations Command, March 20, 2019.

[32] DOE, *Foreign Government Sponsored or Affiliated Activities*, DOE O 486.1A, September 4, 2020; NSF, Office of the Director, *Personnel Policy on Foreign Government Talent Recruitment Programs*, July 11, 2019.

[33] Donald J. Trump, Presidential Memorandum on United States Government-Supported Research and Development National Security Policy, NSPM-33, January 14, 2021.

[34] Council on Governmental Relations, *Framework for Review of Individual Global Engagements in Academic Research: Version 1.0*, January 14, 2020.

[35] Mary Sue Coleman, "Letter to JCORE," Association of American Universities, January 27, 2020.

[36] Michael D. Griffin, *Letter to Academic Colleagues*, U.S. Department of Defense, October 10, 2019.

1982, a National Academy of Engineering (NAE) panel concluded that although "[t]here has been a substantial transfer of U.S. technology—much of it directly relevant to military systems—to the Soviet Union from diverse sources," and Soviet intelligence efforts were targeting U.S. academia, little of the technology transfer occurred through the open communication channels standard for fundamental research.[37] Because additional restrictions could "weaken both military and economic capacities by restricting the mutually beneficial interaction of scientific investigators, inhibiting the flow of research results into military and civilian technology, and lessening the capacity of universities to train advanced researchers" and erode "educational and cultural values," the NAE panel found that for most university activities, the "benefits of total openness overshadow their possible near-term military benefits to the Soviet Union."[38]

In 1985, President Ronald Reagan issued National Security Decision Directive (NSDD) 189, which adopted the main recommendations of the NAE panel as U.S. policy.[39] NSDD 189 directs "that, to the maximum extent possible, the products of fundamental (basic and nonproprietary applied) research remain unrestricted" and that national security classification is the appropriate mechanism to use when control is needed in the interests of national security.[40] NSDD 189 does, however, allow exceptions when needed to be consistent with U.S. law.

While the 1982 NAE panel observed that the PRC had low capacity to apply technologies to military uses and the USG was encouraging Chinese students to attend U.S. universities, the PRC's capabilities have advanced considerably since then.[41] More recently, NSF released a JASON report in 2019 that acknowledged the evolving concerns about undesirable foreign influence, determined that many of those concerns could be addressed in the context of research integrity, and reaffirmed the value of NSDD 189.[42] Broader constraints, the JASON report concluded, would be "neither feasible nor desirable."[43]

---

[37] National Academy of Engineering, *Scientific Communication and National Security*, National Academies Press, 1982, p. 1.

[38] National Academy of Engineering, *Scientific Communication and National Security*, National Academies Press, 1982, pp. 3–4.

[39] Ronald Reagan, *National Policy on the Transfer of Scientific, Technical, and Engineering Information*, NSDD 189, September 21, 1985.

[40] Ronald Reagan, *National Policy on the Transfer of Scientific, Technical, and Engineering Information*, NSDD 189, September 21, 1985, p. 2.

[41] National Academy of Engineering, *Scientific Communication and National Security*, National Academies Press, 1982, pp. 58–59.

[42] JASON, *Fundamental Research Security*, MITRE Corporation, December 2019.

[43] JASON, *Fundamental Research Security*, MITRE Corporation, December 2019, p. 32.

1.3. Project Objectives

In response to concerns that the research strength and practices of strategic competitors, particularly the PRC, threaten the position of the United States as a partner of choice for some foreign researchers, the Basic Research Office (BRO) within the Office of the Under Secretary of Defense for Research and Engineering (OUSD[R&E]) asked the RAND National Defense Research Institute to

1. summarize how DoD approaches IBRC and how DoD fits into the broader research ecosystem[44]
2. identify scientific and strategic considerations for IBRC for DoD
3. formulate suggestions to help DoD improve how it uses IBRC.

1.4. Report Organization

Chapter Two describes the study's methodology. Chapter Three summarizes DoD's IBRC activities and the position of DoD in the basic research ecosystem. In this chapter, the project team compares and contrasts DoD's IBRC approaches with those of four other government basic research funders. Chapter Four then discusses scientific and strategic considerations relevant to IBRC. Chapter Five examines contributors and challenges to IBRC success for DoD. Chapter Six presents observations and suggestions to improve how DoD uses IBRC.

The report also contains four appendixes. Appendix A provides additional information on the four example organizations described in Chapter Three. Appendix B describes the relationship between security cooperation and IBRC at DoD and the implications for the types of collaborations discussed in Chapter Three. Appendix C provides an overview of the scientific and strategic considerations discussed in Chapter Four in the context of two scientific fields: optoelectronics and microelectronics. Appendix D elaborates on the contributors to and inhibitors of international research collaboration success presented in Chapter Five.

[44] The majority of the analysis considered all of DoD, including the Army, the Navy, and the Air Force, collectively. The analysis did not examine the specific roles, responsibilities, and authorities of entities within DoD, such as BRO.

CHAPTER TWO

Methodology

The project team drew on information from both open sources and interviews. This chapter first describes the interview process and how the project team assembled contextual background information. It then describes how the team selected example organizations to compare with DoD and analyzed how those organizations approach IBRC. The chapter ends by identifying some limitations of the work.

2.1. Interviews

The project team conducted two sets of interviews. The first set focused broadly on IBRC. The second set discussed IBRC for microelectronics and optoelectronics but much of the information provided was also applicable to DoD IBRC in general. The project team identified prospective interviewees using suggestions from the sponsoring office, project team members, and other interviewees. The project conducted all interviews on a voluntary basis, obtained informed consent, and agreed not to attribute content to specific individuals. Due to the coronavirus disease 2019 (COVID-19) emergency, the project team conducted all interviews by phone or videoconference. Each interview lasted approximately 60 minutes. Table 2.1 summarizes the organizational affiliations of the interviewees in each set. Two DoD employees participated in both sets of interviews.

The interviews used a semistructured format with interviewers asking follow-up questions based on the initial responses. Interviewers tailored the questions based on the expertise of each interviewee. At least one interviewer and one notetaker attended each interview. The notetaker captured as close to a verbatim record of the discussion as possible.

TABLE 2.1
Interviewee Affiliations

	DoD	USG Other Than DoD	U.S. Academia
Set 1: General IBRC	17	6	0
Set 2: Microelectronics and optoelectronics	9	0	1

The RAND Human Subjects Protection Committee reviewed the project and determined that it was not research. The U.S. Army Medical Research and Development Command, Office of Research Protections, Human Research Protection Office concurred with that determination.

2.1.1. International Basic Research Collaboration Interviews

Between July and October 2020, the project team conducted 16 interviews on IBRC that included a total of 23 people. (Some interviews included up to four interviewees.) All interviewees were federal employees. Seventeen of the interviewees (ten interviews) were DoD civilian or military personnel. The set of DoD interviewees included Army, Navy, and Air Force personnel who were stationed in North America, Europe, Asia, and South America. The remaining interviewees were from NSF, OSTP, DOE, and the Department of State. Interviewees either worked as researchers or worked in offices with international engagement responsibilities. Because the interviews took place before or immediately after DoD finalized its October 2020 *Department of Defense International Science and Technology Engagement Strategy*, none of the interviewees commented on that document or its implementation.[1]

The interviews discussed the following topics about the interviewee's organization:

- types of IBRC activities the organization supports
- reasons the organization supports IBRC
- how the organization measures IBRC success
- how the organization identifies and evaluates IBRC opportunities, including cost, benefit, and risk considerations of both participating and of not participating
- policies, rules, and/or procedures that govern how the organization engages in IBRC
- considerations for proposed IBRC projects with potential for both defense and civilian applications or participants with a mix of defense and civilian connections
- policies, rules, and procedures that govern how the organization engages with non-U.S. persons in the realm of basic research and how researchers that the organization supports interact with non-U.S. researchers and entities based outside of the United States
- experiences, if any, the organization has had with France or China (the countries in which the two non-U.S. example organizations are located: see Section 2.3.2).

Other interview topics included

- challenges and barriers to IBRC
- characteristics of specific S&T topics that affect their suitability for IBRC
- IBRC best practices or lessons learned.

---

1 DoD, *Department of Defense International Science and Technology Engagement Strategy: A Unified Approach to Strengthen Alliances and Attract New Partners*, October 2020.

2.1.2. Microelectronics and Optoelectronics Interviews

Between March and June 2021, the project team conducted ten interviews on IBRC issues relevant to microelectronics and optoelectronics. (Each interview had a single interviewee.) Nine of the interviewees were DoD employees; one worked in academia. The DoD employees worked for the Navy, Army, Air Force, and the Defense Advanced Research Projects Agency (DARPA). Some worked as researchers; others worked in DoD international offices or in program management.

The interviews discussed the following topics:

- basic research focuses for microelectronics and optoelectronics for the interviewee's organization, the USG, and other countries
- expertise and facilities needed for microelectronics and optoelectronics basic research
- distribution of expertise and facilities in the United States and abroad
- landscape of microelectronics and optoelectronics IBRC
- motivations, benefits, risks, and other considerations for IBRC
- differences between costs and benefits of domestic and international collaborations
- challenges and barriers to IBRC, including best practices and lessons learned
- transition paths for microelectronics and optoelectronics technologies in the United States and abroad and associated challenges
- impact of IBRC on transition
- differences between microelectronics and optoelectronics relevant to IBRC and transition
- features of microelectronics and optoelectronics ecosystems that are aligned with U.S. strategic interests and associated policy options.

Information from the interviews specific to microelectronics or optoelectronics is reported in Appendix C. More general information is presented in Chapter Four, particularly Section 4.2.9 on collaboration-related overhead, and in the Chapter Five section on reasons why some DoD researchers are reluctant to participate in IBRC (Section 5.2.3).

2.2. Sources for Information

The project team assembled the information used to provide context and to develop the frameworks using searches in JSTOR, Academic Search Complete, and the National Academies Press catalog. The project team also used sources with which its members were familiar and government-government S&T cooperation agreements available from the U.S. Department of State's "Treaties and Other International Acts Series (TIAS)" webpage (https://www.state.gov/tias/). Additional information about microelectronics and optoelectronics came from analysis of scientific literature and documents published by the European Union (EU), the Congressional Research Service and other USG agencies, professional societies, the commercial sector, and academic organizations. Interviewees also provided background information, in addition to information about specific organizations, including pointers to gray literature (especially from their or other federal government organizations).

Information about the foreign organizations came primarily from the organizations' websites. Other information came from interviewees, news articles, and studies on the countries in which the organizations are located.

Most of the information on the U.S. organizations came from the organizations' public websites and interviews.

2.3. Selection of Example Organizations

In order to understand the range of IBRC approaches in use and their potential relevance to DoD, the project team compared and contrasted the IBRC approaches of other government basic research funders to those of DoD. The team included organizations both inside and outside of the United States in order to provide opportunities to examine both different IBRC approaches within the USG as well as approaches of potential collaboration partners and competitors. While the country in which an organization resides affects its policies and approaches and the team did collect and consider country-level information, the team made organizations, not countries, the primary unit of analysis. Working at the organization level rather than the country level allowed the team to look at within-country differences, such as those that exist between DoD and other USG basic research funders.

As a compromise between breadth and depth, the team decided to include two USG and two foreign organizations in the set of examples. In keeping with the project's focus on basic research, the team selected only organizations with substantial basic research portfolios. To focus on the roles of government in supporting basic research, the team did not include non-government organizations, such as nonprofits or private companies that fund basic research.

The team determined that the organizations in the set should collectively

1. have characteristics analogous to key features of DoD (e.g., mix of applied and basic research, performers, sensitivities such as export and classification controls)[2]
2. highlight differences between DoD and other organizations
3. illustrate exemplary practices
4. cover a broad geographic area
5. support both defense and civilian basic research
6. include an organization in a strategic competitor nation (i.e., Russia or China)
7. have strong connections to the North Atlantic Treaty Organization (NATO) or European Commission-funded research programs (e.g., Horizon 2020).

---

[2] Export control and classification are predominantly concerns for applied, not basic, research.

Additionally, the sponsor suggested that all organizations selected support a mix of scientific disciplines similar to the composition of DoD's basic research portfolio. This guidance was intended to avoid having discipline-specific issues for fields that constitute only a small part of DoD's portfolio, such as the life sciences, dominate the project.

2.3.1. Selection of U.S. Government Example Organizations

As candidate organizations, the team initially considered the 15 USG departments and independent agencies that obligated funds to basic research in fiscal year (FY) 2018 according to the National Center for Science and Engineering Statistics' *Survey of Federal Funds for Research and Development* (Figure 2.1).[3] To satisfy the requirement that the selected organization have a substantial basic research portfolio, the team then narrowed the list of candidates to the top five USG funders of basic research (other than DoD). Each of those organizations has an annual R&D budget in excess of $2 billion and funds roughly $1 billion or more of basic research.

FIGURE 2.1
USG FY 2018 Research Obligations by Organization

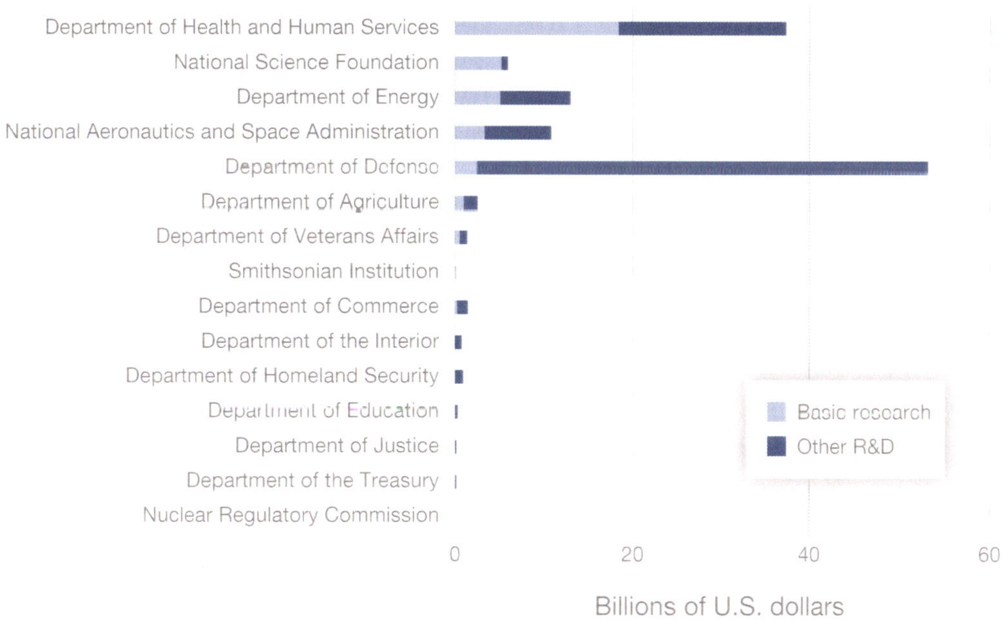

SOURCE: National Center for Science and Engineering Statistics, *Survey of Federal Funds for Research and Development, Fiscal Years 2018–19*, January 2020, Table 5.

NOTES: Shown are the 15 USG departments or independent agencies that obligated funds to basic research in FY 2018. Organizations are ordered by basic research obligations.

[3] National Center for Science and Engineering Statistics, *Survey of Federal Funds for Research and Development, Fiscal Years 2018–19*, January 2020, Table 5.

Next, the team examined the mix of S&T fields each of the candidate organizations supports (Figure 2.2). While HHS and the U.S. Department of Agriculture (USDA) both focus strongly on the life sciences, DOE, NSF, and the National Aeronautics and Space Administration (NASA) have more diverse portfolios that are more similar to DoD's.

The team also examined the mix of performers each of the candidate organizations uses (Figure 2.3). HHS's mix of performers is most similar to DoD's, with both organizations obligating about half of their basic research funds to academia and about 20 percent to intramural laboratories. Like DoD and HHS, NSF also obligates the majority of its basic research funds to universities and colleges. However, unlike DoD, NSF does not have intramural laboratories. DOE has only a single intramural (government-owned and government-operated) laboratory: National Energy Technology Laboratory.[4] While the other 16 DOE national laboratories are federally funded research and development centers (FFRDCs), they serve a function similar to DoD's intramural laboratories and represent the majority of DOE's basic research investments.[5] NASA and DoD rely more heavily on industry for basic research (28 and 15 percent of their portfolios, respectively) than the other organizations shown in Figure 2.3. Uniquely, DoD obligates a larger fraction of its basic research funds to foreign performers (2.8 percent) than any of candidate organizations.[6] HHS, which obligates 0.6 percent

FIGURE 2.2

USG FY 2018 Basic Research Obligations by Field of Science and Organization

Fraction of organization's FY 2018 basic research obligations to the indicated field of science								
	Life sciences	Environmental sciences	Engineering	Computer sciences and mathematics	Other sciences	Psychology	Social sciences	Physical sciences
DOE	0.08	0.06	0.20	0.12	0.10	0.00	0.00	0.43
NSF	0.13	0.23	0.14	0.19	0.07	0.01	0.03	0.20
DoD	0.11	0.04	0.30	0.18	0.19	0.02	0.01	0.15
NASA	0.02	0.34	0.00	0.00	0.00	0.00	0.00	0.64
HHS	0.84	0.01	0.06	0.00	0.03	0.06	0.00	0.00
USDA	0.88	0.01	0.03	0.00	0.00	0.00	0.04	0.05

SOURCE: National Center for Science and Engineering Statistics, *Survey of Federal Funds for Research and Development, Fiscal Years 2018–19*, Table 30.

NOTES: Each cell shows the proportion of its FY 2018 basic research obligations that the indicated organization (row) dedicated to the indicated field of science (column). Shown are the six USG departments or independent agencies that obligated at least $2 billion to R&D in FY 2018. The figure shows organizations with similar obligation profiles and fields to which organizations obligate similar proportions of their funds near each other. The project team used the hierarchical clustering algorithm in Gene Cluster 3.0 (M. J. L. de Hoon, S. Imoto, J. Nolan, and S. Miyano, "Open Source Clustering Software," *Bioinformatics*, Vol. 20, No. 9, 2004) with a city-block distance metric and average linkage to arrange the organizations and fields.

[4] DOE, National Energy Technology Laboratory, "NETL Plays Key Role in DOE's New Lab Partnering Service Web Experience," webpage, July 11, 2018.

[5] NSF, "Master Government List of Federally Funded R&D Centers," webpage, June 2020.

[6] National Center for Science and Engineering Statistics, *Survey of Federal Funds for Research and Development, Fiscal Years 2018–19*, January 2020, Table 27.

FIGURE 2.3
USG FY 2018 Basic Research Obligations by Performer and Organization

Fraction of organization's FY 2018 basic research obligations to the indicated performer							
	Universities and colleges	Intramural	FFRDCs	Industry	Other nonprofits	Foreign	State and local governments
DOE	0.17	0.05	0.71	0.06	0.01	0.00	0.00
NASA	0.19	0.13	0.33	0.28	0.08	0.00	0.00
NSF	0.82	0.00	0.06	0.06	0.06	0.00	0.00
HHS	0.56	0.21	0.02	0.05	0.16	0.01	0.00
DoD	0.54	0.22	0.01	0.15	0.05	0.03	0.00
USDA	0.32	0.66	0.00	0.00	0.01	0.00	0.00

SOURCE: National Center for Science and Engineering Statistics, *Survey of Federal Funds for Research and Development, Fiscal Years 2018–19*, January 2020, Table 27.

NOTES: Each cell shows the proportion of its FY 2018 basic research obligations that the indicated USG organization (row) dedicated to the indicated performer (column). Shown are the six USG departments or independent agencies that obligated at least $2 billion to R&D in FY 2018. The figure shows organizations with similar obligation profiles and performers to which organizations obligate similar proportions of their funds near each other. The project team used the hierarchical clustering algorithm in Gene Cluster 3.0 (M. J. L. de Hoon, S. Imoto, J. Nolan, and S. Miyano, "Open Source Clustering Software," *Bioinformatics*, Vol. 20, No. 9, 2004) with a city-block distance metric and complete linkage to arrange the organizations and performers.

of its basic research funds to foreign performers, has the highest proportion of such obligations among the organizations in the candidate set.[7, 8]

In view of the criteria and the sponsor's preferences, the team selected NSF and DOE as the USG example organizations. Both have substantial basic research portfolios and support a mix of S&T fields similar to those DoD supports. Like DoD, NSF relies heavily on academic researchers. While neither NSF nor DOE has a large intramural basic research enterprise, the tight connections between DOE's national laboratories and DOE make them sufficiently analogous to DoD's intramural laboratories.

NSF and DOE also have some key dissimilarities with DoD. While basic research is the minority of both DOE and DoD's R&D portfolios (39.0 and 4.5 percent, respectively), 87 percent of NSF's R&D portfolio is basic research.[9] Unlike NSF, some of DOE and DoD's later stage research is subject to export and classification controls. Additionally, NSF is widely considered to set the gold standard for funding scientific research, and organizations worldwide have tried to emulate its practices.

---

7 National Center for Science and Engineering Statistics, *Survey of Federal Funds for Research and Development, Fiscal Years 2018–19*, January 2020, Table 27.

8 The amount obligated to foreign performers does not indicate the degree of collaboration or interaction with U.S. researchers that occurred during or as a result of the effort. Additionally, the amounts do not include funds the organization obligated to the U.S. side of IBRCs or funds foreign countries used to support their own researchers collaborating with U.S. researchers as that information was not available.

9 National Center for Science and Engineering Statistics, *Survey of Federal Funds for Research and Development, Fiscal Years 2018–19*, January 2020, Table 5.

2.3.2. Selection of Foreign Government Example Organizations

With countries other than the United States accounting for approximately 75 percent of the world's R&D expenditures in 2017,[10] numerous candidate foreign government organizations exist. The team's initial approach was to assemble high-level information for a broad set of candidate organizations to support the selection effort. However, the team determined that that approach was too resource-intensive because of the lack of data sources that had pre-compiled the desired information at the level of foreign government research funding organizations (vice the level of overall foreign country research funding) and the lack of uniformity in the information that organizations published about themselves.

Thus, the team decided to first select an organization that would satisfy the criterion that one organization be in a strategic competitor country and then to select a second organization that in combination with the first would make the set satisfy all the criteria (e.g., geographic diversity). To further narrow the search set and to ensure that the organizations selected were analogous to multiple aspects of DoD, the team additionally focused on organizations with substantial external funding or policymaking responsibilities, rather than organizations that mainly conducted research.

The *Summary of the 2018 National Defense Strategy of the United States of America* identified China and Russia as strategic competitors.[11] However, China spends substantially more on both basic research and overall R&D than Russia does. For example, China's gross domestic expenditure on basic research in 2019 was 31 billion in 2015 U.S. dollars at PPP, whereas Russia's gross domestic expenditure on basic research in 2016 (the latest year available at the time of the analysis) was about 5.5 billion in 2015 U.S. dollars at PPP.[12] (The United States' gross domestic expenditure on basic research in 2019 was 100 billion in 2015 U.S. dollars at PPP.[13]) In view of China's comparatively large expenditures on R&D and the rate at which China is increasing both its gross domestic expenditures on R&D (12.8 percent from 2010 to 2017[14]) and its share of the world's science and engineering publication output (21 percent in 2018[15]), the team decided to select an example organization in China rather than in Russia.

---

10 NSB, NSF, *Research and Development: U.S. Trends and International Comparisons*, NSB-2020-3, January 15, 2020, pp. 22–23. The publication estimated gross domestic expenditure on R&D in purchasing power parity (PPP) U.S. dollars.

11 Jim Mattis, *Summary of the 2018 National Defense Strategy of the United States of America: Sharpening the American Military's Competitive Edge*, U.S. Department of Defense, undated.

12 OECD, "Gross Domestic Expenditure on R&D by Sector of Performance and Type of R&D," *OECD.Stat*, undated.

13 OECD, "Gross Domestic Expenditure on R&D by Sector of Performance and Type of R&D," *OECD.Stat*, undated.

14 NSB, NSF, *Research and Development: U.S. Trends and International Comparisons*, NSB-2020-3, January 15, 2020, p. 27.

15 NSB, NSF, *Publications Output: U.S. Trends and International Comparisons*, NSB-2020-6, December 17, 2019, p. 5.

As the organization within China on which to focus, the team selected the National Natural Science Foundation of China (NSFC), which is the U.S. NSF's only Chinese counterpart.[16] Other Chinese government organizations the team considered include the Chinese Academy of Sciences (CAS) and the Ministry of Science and Technology (MOST). The team did not select CAS because it is primarily a research institute, not an external research funder. The team did not select MOST, which is NSFC's parent organization, because more public information is available on NSFC than on MOST and because NSFC is more focused on basic research than is MOST.

However, of note, MOST sponsors China's 973 Program, the "National Basic Research Program of China" (国家基础研究计划), which is an academic research program with a strategic objective to

> mobilize China's scientific talents in conducting innovative research on major scientific issues in agriculture, energy, information, resources and environment, population and health, materials, and related areas.[17]

China announced the 973 Program in 1997, and by 1998 it was operational in distributing research grants. According to China's statistical agency, by 2019, MOST had invested approximately 31 billion yuan or approximately 5 billion U.S. dollars over the life of the 973 Program, suggesting an annualized funding level of 1.55 billion yuan or 238 million U.S. dollars per year.[18] In contrast, NSFC's budget in 2019 alone was 31.11 billion yuan.[19]

Having chosen NSFC as one foreign government example organization, the team opted to select a European organization as the second foreign organization to provide geographic balance. Country-wide data from OECD, Eurostat, and Horizon 2020 indicated that the United Kingdom (UK), France, and Germany are Europe's research leaders.[20] The team decided against selecting a UK organization due to concerns that the ongoing changes associated with the UK's exit from the EU would complicate data collection and that the information might rapidly become outdated. When considering Germany and France, the project's sponsor suggested the team select the country that best exemplified the desire of BRO to connect

[16] NSF, Office of International Science and Engineering (OISE), "Counterpart Science Funding Agencies," webpage, undated.

[17] Consulate General of the People's Republic of China in New York, "National Basic Research Program of China (973 Program)," news release, March 5, 2016.

[18] National Bureau of Statistics, People's Republic of China, "Science and Technology Development by Leaps and Innovation Leading a New Chapter—Seventh Series of Reports on Economic and Social Development Achievements in the 70th Anniversary of the Founding of New China [科技发展大跨越 创新引领谱新篇--新中国成立70周年经济社会发展成就系列报告之七]," 2019.

[19] NSFC, "NSFC at a Glance," webpage, undated.

[20] For example, see European Union, "Eurostat," webpage, undated; OECD.Stat, homepage, undated; and Quirin Schiermeier, "Horizon 2020 by the Numbers: How €60 Billion Was Divided up Among Europe's Scientists," *Nature*, December 22, 2020.

DoD researchers to the best in the world regardless of whether those researchers worked for defense or civilian institutions. Because France is substantially more open to looking at defense and civilian research holistically than is Germany,[21] the team decided to use a French organization as the project's final example organization.

Within France, the team considered three options for example organizations: the Agence Nationale de la Recherche (National Research Agency of France) (ANR), the Centre National de la Recherche Scientifique (National Scientific Research Center) (CNRS), and the Direction Générale de l'Armement (Directorate General of Armaments) (DGA). All three organizations have defense connections. ANR partners with DGA on the Accompagnement Spécifique des Travaux de Recherches et d'Innovation Défense (Specific Support for Defense Research and Innovation) (ASTRID) program, which supports dual-use research, including basic and other low technology readiness level (TRL) work, and CNRS accepts ASTRID funds.[22, 23] Like DoD, CNRS has offices abroad.[24] Both CNRS and ANR establish collaboration agreements with international partners, strive to increase ties with other European countries, and are on the U.S. NSF's list of counterparts.[25] CNRS is primarily a research institution, while ANR competitively distributes funding to a range of research performers. Comparatively few public resources are available on DGA. After considering the options, the team selected ANR because it appeared to provide the most public information about its international collaboration processes and approach.

2.4. Scope and Limitations of the Research Approach

This section describes the scope of the project—a high-level overview of DoD IBRC and the associated considerations. The section also describes related limitations. Many aspects of DoD basic research and IBRC, including some described below and in Section 6.2, deserve more-detailed examination than this project was able to provide.

- This report provides high-level information about how the basic research landscape has changed over time (see Section 1.1). However, it does not examine in detail the state of

---

21 See, for example, Christian Mölling, *Defense Innovation and the Future of Transatlantic Strategic Superiority: A German Perspective*, German Marshall Fund of the United States, March 23, 2018.

22 ANR, *Accompagnement Spécifique des Travaux de Recherches et d'Innovation Défense: ASTRID [Specific Support for Defense Research and Innovation Work: ASTRID]*, 2020. France's Agence de l'Innovation de Défense, which is a department attached to DGA, funds the ASTRID program; DGA and DIA both monitor the program.

23 TRLs describe the maturity of a technology and its readiness to be included in an acquisition program. Basic research is the lowest level. DoD, "DAU (Defense Acquisition University) Glossary," webpage, undated.

24 CNRS, "International," webpage, undated.

25 NSF, OISE, "Counterpart Science Funding Agencies," webpage, undated.

basic research in the United States or other countries or the extent to which the investment approaches for basic research the United States or other countries have employed have changed over time. Nor does it examine how collaboration patterns have changed over time.

- This report provides an overview of the approaches DoD uses to identify research strengths in other countries, but the project did not evaluate the effectiveness of those approaches or conduct an independent analysis of the research strengths of the United States or other countries.
- Although the project team spoke with individuals from the Army, the Navy, and the Air Force, it did not conduct detailed analysis of the roles, responsibilities, or authorities of individual DoD entities.
- The study focused on the roles of government organizations, particularly DoD, in IBRC. Due to the emphasis on DoD, most of the interviewees were DoD employees. While the study provides an overview of approaches four example organizations take to IBRC, additional data would be needed to fully describe those organizations' approaches to IBRC as well as the perspectives of nongovernment basic researchers supported by DoD and other organizations.
- The project examined the mechanisms and processes DoD international office persons (IOPs) use to allocate funds and connect their efforts to the DoD research enterprise (see Section 3.1.2.1). It also identified other mechanisms through which DoD and other organizations support IBRC (e.g., joint calls with foreign funders for proposals, allowing some domestic grantees to work with international counterparts who acquire funds independently), but it did not evaluate the relative effectiveness of any of the processes. (However, some interviewees did comment on challenges and opportunities for improvements.) The project also did not consider the optimum amount of funds that might be allocated through each mechanism or the optimum proportion of funds to U.S. and non-U.S. basic research performers.
- Because information on organizations' websites was one of the main data sources used, the content in this report focuses on the image the organizations are trying to project to key audiences, including would-be collaborators. Differences between an organization's public image and its actual practices might exist.
- As other efforts have focused on international students, this project focused on non-U.S. post-graduate researchers who are affiliated with organizations in their home countries. While trends in the quality and availability of higher education in the United States and abroad are relevant for the DoD workforce and the vitality of the DoD basic research enterprise, the authors did not examine these factors.
- A number of topics, while largely outside of the scope of this project, could be fruitful areas for future studies. Suggestion 5 in Section 6.2, which discusses opportunities for DoD to assemble a fuller picture of its IBRC efforts and processes, elaborates on some of the items below.

- In order to focus on the strategic and scientific considerations most relevant to IBRC (see Chapter Four), the project did not examine the strategic and scientific considerations that influence overall organizational decisions about the research fields in which to invest (independent of where the research is performed). Similarly, the project did not determine how DoD weighs those considerations or the potential impact of different prioritizations of the considerations.
- While the project discusses the intent of DoD to connect its IBRC efforts with the DoD intramural research enterprise (see Sections 3.2.4 and 5.1), the project did not examine the effectiveness of those efforts or the relationship between DoD IBRCs and the rest of the DoD basic research portfolio (e.g., extent to which IBRC complements other efforts, relative value of IBRC and non-IBRC efforts). The *Department of Defense International Science and Technology Engagement Strategy* indicates that DoD should optimize its international S&T portfolio to fill gaps in the domestic S&T portfolio and address regional priorities, and it outlines an approach that the services can use to combine that information to formulate international S&T engagement plans.[26]
- As discussed in Sections 4.2.5 and 4.2.6, awareness of the potential applications of an IBRC project can impact the interest of DoD and partners in pursuing it. However, the project did not comprehensively examine the challenges with continuing a research international collaboration after the work progresses beyond the fundamental research stage[27] or whether changes to how DoD approaches such later-stage international collaborations might impact DoD basic research or the overall DoD research ecosystem.
- While the report includes information about USG obligations to foreign basic research performers (see Table 3.4), the project did not assemble complete information about how much of DoD's budget is used to support IBRC. As discussed in Section 6.2 (Suggestion 5), data are not readily available on (1) funds that USG entities have awarded to U.S. researchers through bilateral or multilateral calls for IBRC projects and (2) funds that support the work of U.S. researchers who develop bottom-up collaborations with foreign researchers who acquire funding independently. Additionally, other DoD support for IBRC comes from federal employees who spend a portion of their time on activities that enable IBRC (e.g., developing policies, evaluating proposals, conducting security reviews).

---

[26] DoD, *Department of Defense International Science and Technology Engagement Strategy: A Unified Approach to Strengthen Alliances and Attract New Partners*, October 2020.

[27] NSDD 189 applies only to fundamental research, and, unlike fundamental research, the results of non-fundamental research are not ordinarily published openly (Ronald Reagan, *National Policy on the Transfer of Scientific, Technical, and Engineering Information*, National Security Decision Directive 189, September 21, 1985).

CHAPTER THREE

Position of the U.S. Department of Defense in the Basic Research Ecosystem

This chapter first provides an overview of how DoD approaches IBRC, including key organizations and activities. Then it compares and contrasts DoD's IBRC approaches to those of four example organizations and reflects on the connections between DoD's approaches and its mission.

3.1. Overview of Department of Defense Approaches to International Basic Research Collaborations

DoD invests in basic research to encourage "advances in fields that are likely to contribute to national defense" and to "foster a competitive technology base for the U.S. military."[1] In FY 2018, DoD (including the defense agencies and the services) obligated $2.4 billion to basic research, which was 4.5 percent of DoD's R&D obligations.[2] As shown in the previous chapter (Figure 2.3), DoD's main basic research performers are universities and colleges (54 percent), its own intramural laboratories (22 percent), industry (15 percent), nonprofits other than FFRDCs (5 percent), and foreign entities (3 percent).[3]

Some characteristics of basic research create unique implications for DoD IBRC compared with the later stage research that dominates DoD's R&D portfolio. Thus, some items that feature prominently in the 2020 *Department of Defense International Science and Technology Engagement Strategy* typically have low direct relevance to IBRC, although they may be indirectly applicable though the impact of preparations to transition results of basic

---

1 National Research Council, *Assessment of Department of Defense Basic Research*, National Academies Press, 2005, p. ix.

2 National Center for Science and Engineering Statistics, *Survey of Federal Funds for Research and Development, Fiscal Years 2018–19*, January 2020, Table 5. FY 2018 was the most recent year available at the time of the analysis.

3 National Center for Science and Engineering Statistics, *Survey of Federal Funds for Research and Development, Fiscal Years 2018–19*, January 2020, Table 30.

research to higher TRLs.[4] While the strategy's vision includes a portfolio of engagements that is "integrated into, and benefits, funded programs of record," basic research, by definition, does not address a specific application.[5] The strategy focuses on government-to-government cooperation, which "is often the most appropriate route" for higher-TRL defense efforts; as discussed in Section 3.1.2, many IBRCs do not employ the government-to-government paradigm.[6] The strategy indicates that international agreements, which can be time-consuming to establish, are often a key consideration when considering an international S&T cooperative opportunity. In contrast, international agreements are rarely needed for DoD IBRC (see Appendix B). The strategy identifies the need to "protect the security of critical U.S. technologies" as one of the principles of international S&T cooperation and indicates that the Strategic Technology Protection and Exploitation directorate that is part of OUSD(R&E) provides guidance on protecting specific technologies.[7] While the need to protect critical technologies is a consideration for IBRC, the results of basic research are normally published openly, and export controls rarely apply (see Section 4.2.6).

3.1.1. Key Organizations

Multiple organizations within DoD have responsibilities for basic research.[8] For example,

- **BRO within OUSD(R&E)** provides strategic guidance and coordination to the DoD basic research enterprise.[9] BRO also makes investments in areas in which the services might not be able.[10] BRO's goals include providing direction, coordination, and oversight; improving the science and engineering workforce; enhancing university-industry collaborations relevant to DoD; and engaging with academic and international partners.[11]

---

4 DoD, *Department of Defense International Science and Technology Engagement Strategy: A Unified Approach to Strengthen Alliances and Attract New Partners*, October 2020.

5 DoD, *Department of Defense International Science and Technology Engagement Strategy: A Unified Approach to Strengthen Alliances and Attract New Partners*, October 2020, p. 6.

6 DoD, *Department of Defense International Science and Technology Engagement Strategy: A Unified Approach to Strengthen Alliances and Attract New Partners*, October 2020, p. 13.

7 DoD, *Department of Defense International Science and Technology Engagement Strategy: A Unified Approach to Strengthen Alliances and Attract New Partners*, October 2020, pp. 3, 7.

8 Documenting all entities with basic research or international responsibilities and their associated chains of command is outside the scope of this report.

9 DoD, BRO, "Our Mission," webpage, undated.

10 DoD, BRO, "Our Mission," webpage, undated.

11 DoD, BRO, "Our Mission," webpage, undated.

- The **Office of Naval Research (ONR)** manages the full spectrum of U.S. Navy research, including basic research, applied research, and advanced technology development. Within ONR, ONR Global focuses on international connections.
- The **Air Force Office of Scientific Research (AFOSR)** manages the U.S. Air Force's basic research portfolio. The Air Force Office of Scientific Research International Office (AFOSR/IO) connects the U.S. Air Force to overseas basic research efforts.
- The **U.S. Army Research Office (ARO)**, which is part of the U.S. Army Combat Capabilities Development Command (CCDC), manages the U.S. Army's extramural basic research portfolio.
- **DARPA** seeks to fund transformational, rather than incremental, research, including basic research.
- The Army, the Navy, the Air Force, and the Office the Secretary of Defense have **intramural laboratories** that work on the full spectrum of defense research, including basic research.

AFOSR/IO, ONR Global, and CCDC all station personnel around the world. The personnel have a range of titles (e.g., science director, international program manager, technical director), and for clarity and simplicity this report refers to the group collectively as international office persons (IOPs). IOPs are responsible for understanding the state of research in a geographic region and on multiple scientific topics. IOPs merge that knowledge with their understanding of their services' current and future R&D activities and needs across the spectrum of TRLs to identify places where connecting international expertise to their services' research enterprise would be beneficial. As discussed later in this chapter, the global presence of IOPs is a unique and prominent feature of DoD's approach to international basic research engagement. Because DoD IOPs are the face of DoD basic research to many around the world, the project examined their functions in comparative depth. The project collected comparatively less data on other ways in which DoD engages in and supports IBRC and did not attempt to evaluate the relative effectiveness of the different approaches.

IOPs also coordinate with their services' international offices (i.e., Navy International Programs Office, Air Force International Affairs, Office of the Deputy Assistant Secretary of the Army for Defense Exports and Cooperation). The services' international offices manage their services' overall relationship with foreign counterparts, including cooperative development and security assistance programs. (See Appendix B for a primer on the distinctions between security cooperation and IBRC.)

The Office of the Secretary of Defense (OSD) ensures that DoD provides consistent messaging to international partners.[12] The Under Secretary of Defense for Research and Engineering represents DoD on international S&T matters and provides oversight.[13] The 2020

[12] DoD, *Department of Defense International Science and Technology Engagement Strategy: A Unified Approach to Strengthen Alliances and Attract New Partners*, October 2020.

[13] DoD, *Department of Defense Directive 5137.02: Under Secretary of Defense for Research and Engineering (USD[R&E])*, July 15, 2020.

Department of Defense International Science and Technology Engagement Strategy provides additional information on the DoD entities with responsibilities for international S&T cooperation.[14]

3.1.2. Key Activities

DoD uses a range of mechanisms to facilitate and support IBRC that are discussed below in the context of the framework described in Table 1.1.

3.1.2.1. Funding Research or Exploratory Activities with an International Component

DoD IOPs interviewed for the project indicated that they conduct multiple types of exploratory activities to understand the international research ecosystem and identify funding opportunities including reading the literature, attending conferences, exchanging information with the U.S. defense industry about its international basic research portfolio, and arranging (pre-COVID-19) liaison visits to laboratories of foreign principal investigators (PIs). Multiple interviewees emphasized the importance of liaison visits for their ability to facilitate building networks, to learn the ground truth about capabilities, and to keep more current than would be possible by relying on publications alone. However, DoD IOPs are not solely responsible for collecting and analyzing information to understand the international basic research landscape (see Chapter Five).

DoD IOPs also fund a range of international basic research or supporting exploratory activities (see Table 3.1).

DoD and foreign government partners (not necessarily ministries of defense) issue joint calls to co-sponsor research projects that include team members from each country (see Table 3.2). In joint calls, each research team typically submits a single joint proposal, but each sponsor funds the researchers from its country.

DoD entities also run several international exchange programs:

- The Army, the Navy, and the Air Force participate in the **Engineer and Scientist Exchange Program (ESEP)**, which allows DoD and foreign ministries of defense to place military and civilian researchers in each other's research, development, test, and evaluation facilities to conduct long-term (e.g., one or more years) work.[15, 16] The ESEP program includes only partners and allies with whom DoD has signed a bilateral

[14] DoD, *Department of Defense International Science and Technology Engagement Strategy: A Unified Approach to Strengthen Alliances and Attract New Partners*, October 2020, pp. 9–10.

[15] Office of the Director for International Cooperation, Office of the Under Secretary of Defense for Acquisition, Technology and Logistics, *International Cooperation in Acquisition, Technology and Logistics (IC in AT&L) Handbook*, 7th ed., May 2012.

[16] U.S. Air Force Research Laboratory (AFRL), AFOSR, *AFOSR International Office FY19 Annual Report*, undated, pp. 47–48.

TABLE 3.1

Types of International Research and Exploratory Activities That DoD International Office Persons Fund

Mechanism	AFOSR/IO	CCDC	ONR Global
Grants to international PIs through their services' long-range broad agency announcements using funds dedicated to that purpose	Yes	Yes	Yes
Cooperative agreements with international PIs	No	Yes	No
Travel for foreign PIs to meet service personnel at conferences or service laboratories[a]	Yes (Windows on Science)	Yes (Visiting Scientist Program)	Yes (Visiting Scientist Program[b])
Support for international workshops and conferences	Yes (Conference and Workshop Support Program)	Yes	Yes (Collaborative Science Program)
Placements (weeks or months) for service research personnel in foreign, nongovernment laboratories	Yes (Windows on the World)[c]	No	No

[a] Additionally, BRO includes foreign participants in its Future Directions Workshops series.

[b] ONR, "ONR Global Funding Opportunities," webpage, undated.

[c] AFRL, AFOSR, *AFOSR International Office FY19 Annual Report*, undated, p. 6.

TABLE 3.2

Example DoD Joint Basic Research Funding Opportunities with Foreign Partners

Mechanism	DoD Funder(s)	DoD Manager(s)	DoD Funds U.S. Side only?	Partner Countries
Multidisciplinary University Research Initiative	ARO, AFOSR, ONR	ARO, AFOSR, ONR	Yes	Varies (UK and Australia have participated in select topics.)
Bilateral Academic Research Initiative	OSD	ARO, AFOSR, ONR	Yes	UK[a]
Broad agency announcements amplifications	AFOSR	AFOSR	Yes	Varies (Recent partners include the Republic of Korea and Taiwan.[b])
Global Initiatives Program	AFOSR	AFOSR	Depends	Varies (Recent partners include Australia, Israel, Republic of Korea, Mexico, and Taiwan.[c])

NOTE: IOP interviewees indicated that the Global Initiatives Program takes multiple forms. In some cases, AFOSR and the foreign partner issue a joint call for projects with research teams from the two countries and each organization funds its country's part of the teams. In others, AFOSR and the foreign partner both award grants to researchers in the partner country on a topic of mutual interest.

[a] DoD, "DoD Announces BARI Award for US-UK Collaboration on Human-Machine Teaming," press release, September 25, 2018.

[b] AFOSR and the National Research Foundation of Korea, "AFOSR—NRF Korea Quantum Information Science and Technologies 2020–2022: Amplification of AFOSR BAA—FA9550-19-S-0003," undated; AFOSR, "MOST-AFOSR Taiwan Topological and Nanostructured Materials," *All Partners Access Network*, undated.

[c] AFRL, AFOSR, *AFOSR International Office FY19 Annual Report*, undated, pp. 95–120.

ESEP Memorandum of Understanding (MOU). While most ESEP participants work on applied topics, basic research work is possible.

- The **International Supplemental Student Exchange Program (ISEP)** funds graduate students of AFOSR-funded PIs to work with an overseas collaborator and allows graduate students of overseas collaborators of AFOSR-funded PIs to work with the AFOSR-funded PI, which opens up the U.S. research enterprise to international collaboration. While most of the exchanges involve academic laboratories, in FY 2019, one student came to AFRL.[17]

3.1.2.2. Permitting and Enabling the National Research Enterprise to Engage Internationally

There are multiple ways that DoD allows international researchers to access academic and service laboratories within the U.S. research enterprise. For example, one DoD interviewee indicated that many intramural DoD researchers collaborate with international colleagues, in many cases continuing relationships that started during graduate school. University personnel, including non-U.S. citizens, who are working on a DoD grant may access DoD's high-performance computing resources.[18] Some programs, such as the Multidisciplinary University Research Initiative program, that prohibit foreign entities from receiving program funds either directly or through subgrants still allow foreign entities to collaborate using other sources of funds.[19] Other programs, such as the Minerva Research Initiative (a university-based social science research program), have allowed foreign universities to submit applications.[20] ARO's Short-Term Innovative Research program is open to both foreign organizations and foreign public entities.[21]

Additionally, the services have several programs that, while not focused on foreign researchers, can be used by visiting foreign researchers to access DoD intramural laboratories. For example, through the Open Campus program, the Army opens parts of its Adelphi Laboratory to visiting researchers from academia and industry, including foreign nationals.[22]

---

[17] AFRL, AFOSR, *AFOSR International Office FY19 Annual Report*, undated, pp. 50–51.

[18] DoD High-Performance Computing Modernization Program, *HPCMP New Users' Guide: "Who May Run on HPCMP Resources?"* undated.

[19] ONR, ARO, and AFOSR, "Fiscal Year (FY) 2020 Department of Defense Multidisciplinary Research Program of the University Research Initiative," undated, pp. 7–8. While, as described above, some Multidisciplinary University Research Initiative calls have included prearranged mechanisms for UK or Australian researchers to obtain money from their governments to participate in select topics, the program also allows bottom-up collaborations.

[20] DoD, Washington Headquarters Services, Acquisition Directorate, *Funding Opportunity Announcement #WHS-AD-FOA-18: Minerva Research Initiative*, undated, p. 7.

[21] ARO, *2019 ARO in Review*, U.S. Army Combat Capabilities Development Command, Army Research Laboratory, undated, p. 391.

[22] U.S. Army, Army Research Laboratory, "Partnering Mechanisms | Open Campus," webpage, undated.

Similarly, some of the placements in Army research laboratories and the Naval Postgraduate School available through the National Research Council (NRC) Resident Research Associateships program are open to non-U.S. citizens.[23] Table 3.3 summarizes these programs alongside the programs dedicated to international exchanges discussed above.

3.1.2.3. Working with Partner Governments and International Organizations to Facilitate IBRC

DoD interviewees indicated that while DoD IOPs spend the bulk of their time engaging the academic community, they also engage with officials from partner governments and international organizations (one to two interviewees identified each activity):

- DoD IOPs liaise with foreign government personnel to share information about basic research investments and inquire about interest in coordinating funding, such as through a joint call.
- Some DoD IOPs also coordinate with the NATO Science and Technology Organization.
- DoD IOPs often coordinate with and support bilateral military-to-military engagements, such as those led by the Office of the Under Secretary of Defense for Acquisition and Sustainment.

DoD also works with partner organizations to establish joint funding mechanisms, such as for the programs described in Table 3.2. DoD partners with both civilian organizations

TABLE 3.3
Programs That Open DoD Service Laboratories to Visiting Foreign Researchers

Mechanism	Dedicated to International Exchanges?	Air Force	Army	Navy
Open Campus Initiative	No	No	Yes	Yes
NRC Resident Research Associateships	No	No	Yes	Yes
ESEP	Yes	Yes	Yes	Yes
ISEP	Yes	Occasional	No	No

NOTE: The ISEP program funds graduate students of AFOSR-funded PIs to work with an overseas collaborator and allows graduate students of overseas collaborators of AFOSR-funded PIs to work with the AFOSR-funded PI. While most of the exchanges involve academic laboratories, in FY 2019, one student came to AFRL (AFRL, AFOSR, *AFOSR International Office FY19 Annual Report*, undated, pp. 50–51). While all the programs contain basic research, some (e.g., ESEP) also include applied work.

23 National Academy of Sciences, National Academy of Engineering, and Institute of Medicine, "Research Associateship Programs: Participating Agencies," webpage, undated.

(e.g., National Research Foundation of Korea) and defense organizations (e.g., UK Ministry of Defence).[24, 25]

3.2. Comparisons Between How the Department of Defense and Other Organizations Approach International Basic Research Collaboration

While no other organization is identical to DoD, DoD shares some similarities with other organizations. In order to understand the variety of IBRC approaches in use and their potential relevancy to DoD, the project team examined the IBRC efforts of several other government basic research organizations—in the United States and overseas—in addition to DoD. Comparing and contrasting the IBRC approaches of other organizations also helped clarify how DoD is positioned within the basic research ecosystem and the extent to which it has adapted its IBRC approaches to its mission.

3.2.1. Overview of Example Organizations

The project team examined four example government basic research funding organizations, two inside the United States and two outside of the United States:

- DOE
- NSF
- ANR
- NSFC.

Table 3.4 summarizes key information about DoD and the organizations. See Chapter Two for information about the process used to select the organizations and Appendix A for additional information on each organization. All of the four organizations and DoD support research on a broad set of topics, and all four of the example organizations devote larger shares of their R&D budgets to basic research than does DoD.

All four of the example organizations almost exclusively fund only their own countries' participation in basic research. In contrast, DoD obligates a comparatively large share of its basic research budget—2.84 percent—to foreign entities.[26] Basic research funding to foreign

[24] DoD, "DoD Announces BARI Award for US-UK Collaboration on Human-Machine Teaming," press release, September 25, 2018.

[25] AFOSR and the National Research Foundation of Korea, "AFOSR—NRF Korea Quantum Information Science and Technologies 2020–2022: Amplification of AFOSR BAA - FA9550-19-S-0003," undated.

[26] National Center for Science and Engineering Statistics, *Survey of Federal Funds for Research and Development, Fiscal Years 2018–19*, January 2020, Table 27.

TABLE 3.4

Key Statistics for the Department of Defense and Selected Example Organizations

Organization Attribute	ANR	DoD	DOE	NSF	NSFC
Country	France	United States	United States	United States	China
R&D budget[a]	€765 million ($666 million)	$53.0 billion	$12.8 billion	$5.8 billion	31.11 billion yuan ($4.52 billion)
Basic research portion of R&D budget	Mix of basic and applied research	4.5% ($2.4 billion)	39.0% ($5.0 billion)	87% ($5.1 billion)	Mainly basic
Basic research budget portion to foreign performers[a, b]	None	2.84% ($68.0 million)	0.01% ($0.6 million)	0.40% ($20.5 million)	Unknown, but likely low
Topical focus	Broad	Broad	Broad	Broad	Broad
Tightly associated laboratories[c]	No	Yes	Yes	No	No
Mission-focused[d]	No	Yes	Yes	No	No

SOURCES: National Center for Science and Engineering Statistics, *Survey of Federal Funds for Research and Development, Fiscal Years 2018–19*, January 2020, Tables 5, 27; ANR, *Soutenir la recherche dans sa diversité—Funding Research in All Its Diversity: Rapport d'Activité 2019—Annual Report 2019*, undated, p. 72; NSFC, "NSFC at a Glance," webpage, undated; and RAND analysis.

NOTE: Data are for the indicated organizations, not the organizations' countries.

[a] Numbers for U.S. organizations are from FY 2018. ANR and NSFC numbers are from 2019. Conversions to U.S. dollars used exchange rates on December 31, 2018, per Exchange-Rates.org.

[b] Numbers exclude funds obligated to researchers from the organization's country working on international projects as those data were not available.

[c] This includes DoD's intramural laboratories as well as the DOE national laboratories, most of which are government-owned and contractor-operated. NSF's FFRDCs are comparatively less tightly associated with NSF.

[d] An organization is mission-focused if it conducts or supports basic research primarily for its own benefit.

entities is just one indicator of the level of international engagement. Another is the amount of funds they obligate to researchers in their own countries who work on international collaborations, but few organizations publish complete data on that metric. Both DOE and DoD are unique in being mission-focused, in that they conduct and support basic research largely for their own organizational benefit. (Because basic research supports many applications, basic research investments also generally benefit society broadly.) As discussed below, many aspects of how DoD conducts IBRC—including directly supporting foreign researchers—seek to support DoD's mission by connecting researchers and transitioning basic research discoveries into the U.S. defense research enterprise.

Tables 3.5, 3.6, and 3.7 summarize some of the ways the example organizations support and facilitate IBRC using the categories described in Table 1.1. More-detailed information is in Section 3.1 and Appendix A.

TABLE 3.5

Types of Research or Exploratory Activities with an International Component That the Example Organizations and DoD Fund

Organization	Examples
ANR	• Runs program to fund French side of collaborations that occur under MOUs with foreign funding counterparts or through European Initiatives (e.g., European Research Area Network) • Funds efforts to help French researchers build networks in an effort to increase their chances of submitting successful proposals in response to international calls
DoD	• Stations IOPs abroad to conduct multiple types of exploratory activities, including attending conferences and visiting laboratories of foreign academic PIs. • Issues joint calls with foreign government partners (not necessarily ministries of defense) to co-sponsor research projects that include team members from each country • Runs international researcher exchange programs
DOE	• Supports international collaborations on large-scale scientific infrastructure, both within the United States and abroad • Runs joint funding calls with counterpart funding agencies
NSF	• Issues joint calls with foreign funding agencies for teams of U.S. and foreign researchers • Funds activities intended to strengthen international networks (e.g., sharing resources, knowledge, and expertise) • Sends NSF experts abroad to build networks and explore collaboration opportunities • Participates in international collaborations that support large-scale international research facilities
NSFC	• Runs programs to fund the Chinese side of projects that take place under bilateral and multilateral MOUs and those that occur without an MOU between funding organizations • Supports personnel exchanges and conferences that take place under bilateral and multilateral MOUs • Funds non-Chinese citizens who are less than 40 years old to conduct research in China

NOTE: See Sections 3.1.2.1, A.1.1.1, A.2.1.1, A.3.1.1, and A.4.1.1 for examples and sources. The project did not seek to assemble a comprehensive list.

Table 3.5 provides an overview of some of the research or exploratory activities with a mandatory international dimension that DoD and the four example organizations fund. All of the organizations in the set issue joint calls for proposals with foreign funding counterparts. NSFC offers grants for non-Chinese citizens who are less than 40 years old to conduct research in China.[27] ANR and NSF sponsor efforts to help domestic researchers build networks with international colleagues.

Table 3.6 provides an overview of some of the ways that DoD and the four example organizations permit and enable their national research enterprises to engage internationally. Unlike the activities in Table 3.5, international engagement is not mandatory for the activities in this category. A common activity in this category is to allow, but not require, some grant recipients to work with foreign partners that have their own funds. Allowing non-domestic

[27] NSFC, *National Natural Science Fund Guide to Programs 2019*, English version, undated, p. 218.

TABLE 3.6

Ways the Example Organizations and DoD Permit and Enable Their National Research Enterprises to Engage Internationally

Organization	Examples
ANR	• Provides some funding opportunities that allow, but do not require, foreign partners that have their own funds • Administers calls for other French funders that allow, but do not require, foreign partners that have their own funds
DoD	• Permits intramural DoD researchers to collaborate with international colleagues • Allows university personnel, including non-U.S. citizens, working on a DoD grant to access DoD's high-performance computing resources[a] • Allows foreign universities to apply to some grant programs • Allows some funding recipients to collaborate with foreign entities who have their own funds • Opens select intramural laboratories and research positions to non-U.S. citizens
DOE	• Opens some DOE national laboratory user facilities to international researchers • Allows DOE national laboratories to enter into some agreements to provide expertise to international and U.S. companies, nonprofit organizations, academia, and state or local governments • Opens some funding opportunities to non-domestic applicants who can demonstrate expertise or capabilities not available domestically
NSF	• Allows researchers applying for funds to include foreign collaborators and international travel and activities in their proposals (NSF normally funds only the U.S. side) • Allows national facilities that NSF manages through cooperative agreements to provide opportunities for foreign researchers to use their facilities • Allows foreign entities to participate in NSF centers using non-NSF funding
NSFC	• Does not appear to provide funding opportunities that allow, but do not require, foreign partners[b]

NOTE: See Sections 3.1.2.2, A.1.1.2, A.2.1.2, A.3.1.2, and A.4.1.2 for examples and sources. The project did not seek to assemble a comprehensive list.

[a] DoD High-Performance Computing Modernization Program, *HPCMP New Users' Guide: "Who May Run on HPCMP Resources?"* undated.

[b] NSFC appears to have separate programs for funding (1) domestic-only projects, (2) international collaborations that take place under an MOU between funding organizations, and (3) international collaborations that take place outside of such an MOU. In contrast, many other organizations consider proposals for international collaborations that take place outside of an MOU as part of calls that also accept proposals for domestic-only projects. Because categories (2) and (3) for NSFC have an obligatory international component, they are listed in Table 3.5.

researchers to apply for grants, as DOE and DoD sometimes do, also fits in this category. Permitting foreign researchers access to some intramural facilities, as DoD and DOE do, or allowing foreign researchers to use national facilities managed under cooperative agreements, are other activities in the category.

Table 3.7 provides an overview of how DoD and the four example organizations work with partner governments and international organizations to facilitate IBRC. Each works with counterpart funding organizations to create vehicles for joint funding calls (see Section 3.1 and Appendix A for additional details). Other common activities in this category include supporting international working groups and meetings. Based on its published materials, ANR appears to puts the most emphasis on coordinating its research priorities with those of

TABLE 3.7
Ways That the Example Organizations and DoD Work with Partner Governments and International Organizations to Facilitate IBRC

Organization	Examples
ANR	• Establishes cooperative agreements with counterparts to provide frameworks and mechanisms for research cooperation • Connects its research priorities to those of other European and international entities
DoD	• Engages with officials from partner government and international organizations to share information about basic research investments and to support a range of military-to-military engagements • Works with partner organizations to establish joint funding mechanisms
DOE	• Stations some employees abroad, usually at U.S. embassies, to support DOE activities • Establishes cooperation agreements with foreign government entities, including agreements that allow DOE labs to share and receive non-public data with foreign partners or allow U.S. researchers to work in foreign facilities • Participates in international working groups, international partnerships, and many bilateral dialogs
NSFC	• Establishes cooperative agreements with counterparts to provide frameworks and mechanisms for research cooperation • Participates in bilateral and multilateral leadership meetings with counterparts
NSF	• Establishes joint funding agreements with foreign government and international organizations • Participates in international forums

NOTE: See Sections 3.1.2.3, A.1.1.3, A.2.1.3, A.3.1.3, and A.4.1.3 for examples and sources. The project did not seek to assemble a comprehensive list.

other governments and international organizations, although other organizations in the set likely also conduct this activity.[28]

3.2.2. Funding Mechanisms

The organizations that the project team examined differ in the IBRC funding approaches they use and prefer. Among the wide variety of funding approaches the organizations use, the project team identified four major archetypes that are shown in Figure 3.1. Some, such as direct funding of foreign researchers, are top-down approaches, with funders often identifying candidate researchers and helping to scope research topics. Others, such as uncoordinated funding for joint projects, are largely bottom-up, with researchers exercising considerable autonomy in selecting partners, topics, and collaboration terms. Coordinated funding for joint projects falls toward the middle of the top-down/bottom-up continuum. Program managers define the overarching collaboration terms and possible topics and countries; subject to those constraints, researchers choose the specific topics and partner. While the archetypes do not include all possible funding arrangements, they do illustrate common ways of

[28] For example, see ANR, *Work Programme 2021*, English version, v.1.1, October 22, 2020, p. 5.

FIGURE 3.1

Common Government Approaches to Funding IBRC Activities Shown from the U.S. Point of View

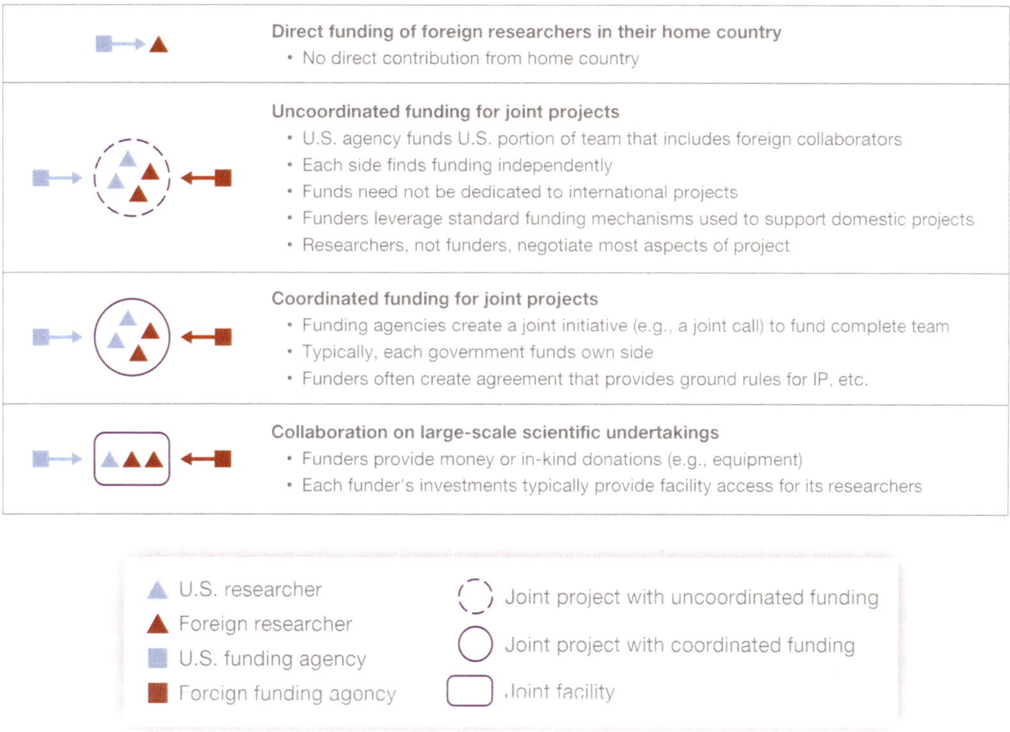

NOTES: Many variants of each type exist, and the archetypes do not include all possible funding arrangements. Multiple foreign agencies and researchers can contribute to a single activity.

structuring support for international research. For example, mechanisms not shown include personnel exchange programs and allowing foreign researchers to access domestic facilities.

Figure 3.2 shows the team's assessment of the extent to which DoD and the example organizations use each of the funding archetypes. The assessments are subjective and are based on the information that the team obtained about each organization.[29] While the discussion includes information interviewees provided about why their organizations prioritize certain archetypes, the project did not assess whether a different weighting of archetypes would help the organizations meet their goals. To compare organizational preferences, intensities of effort, and institutional priorities (but not the level of funds invested through each mechanism), the team assigned one of three levels to each organization for each archetype:

[29] Information on DoD primarily came from interviews with IOPs.

FIGURE 3.2
Funding Archetypes That DoD and Example Organizations Use

	Archetype	NSF	DOE	DoD	ANR	NSFC
	Direct funding of foreign researchers	Uncommon	Uncommon	Primary	Uncommon	Uncommon
	Uncoordinated funding for joint projects	Primary	Primary	Secondary	Secondary	Secondary
	Coordinated funding for joint projects (e.g., joint call)	Primary	Secondary	Secondary	Primary	Primary
	Collaboration on large-scale scientific undertakings	Primary	Primary	Uncommon	Uncommon	Uncommon

U.S. researcher
Foreign researcher
U.S. funding agency
Foreign funding agency
Joint project with uncoordinated funding
Joint project with coordinated funding
Joint facility

NOTES: The indicated preferences do not provide information about the amount of funds that the organizations invest through each mechanism. Data on the total amount of funds that each organization provides through each mechanism were not available.

- **Primary** refers to archetypes that an organization uses commonly as well as those mechanisms that are focal points of an organization's international outreach efforts. A rating of primary does not provide information about the amount of funds the organization invests through the mechanism.
- **Secondary** refers to archetypes that an organization uses that are either less common or are not among the mechanisms that interviewees or organizational materials identified as an organization's emphases. A rating of secondary is not intended to suggest a lack of capacity or engagement but rather is a comparative assessment intended to highlight differences among the organizations.
- **Uncommon** refers to archetypes that an organization uses rarely or never.

Among the organizations examined, DoD appears unique in its emphasis on directly funding foreign researchers in foreign institutions in the absence of a project with domestic researchers.[30,31] DoD IOPs control some of these funds to foreign researchers, but some other DoD basic research programs also allow foreign performers (see Section 3.1.2). Interviewees indicated that the approach is part of a long-term strategy to develop relationships with promising researchers, usually early in their careers, and ultimately integrate them into the DoD research enterprise, which is consistent with DoD being a mission-focused organization.[32] A DoD interviewee indicated that DoD's ability to directly fund foreign researchers allows DoD to act unilaterally, which can provide flexibility and agility. (Bottom-up approaches, such as the uncoordinated funding for joint project archetypes where teams self-organize and each partner finds funding independently can also provide flexibility and agility.) While China also directly funds foreign researchers abroad (e.g., foreign talent recruitment programs), NSFC does not appear to directly provide funds for this purpose.[33]

In contrast, all five organizations fund collaborations with foreign researchers both through uncoordinated and coordinated joint project mechanisms. As these approaches need not directly fund non-domestic researchers, there are fewer political sensitivities to navigate. Among the organizations in the set, only NSFC published the numbers of collaborations of each type that it supported. In 2018, NSFC funded 106 projects in the uncoordinated joint funding category and 324 in the coordinated joint funding category.[34] To convey the difference in volume, the project team considered coordinated funding to be a primary mechanism for NSFC and uncoordinated funding a secondary mechanism. The team considered coordinated and uncoordinated funding of joint projects primary and secondary mechanisms, respectively, for ANR due to the emphasis that ANR puts on agreements that support coordinated funding of projects in its published materials. Due to the emphasis DoD interviewees placed on direct funding of foreign researchers and the prominence of the mechanism in the services' broad agency announcements and DoD international office websites, the team considered funding joint projects through either coordinated or uncoordinated mechanisms comparatively secondary.[35] Because one DOE interviewee indicated that setting up coordinated funding for investigator-initiated projects was a comparatively low priority, the team

---

30 Grants and contracts for foreign scientific research should support specific DoD requirements and be for work that either is not or could not be conducted at a U.S. institution. See U.S. House Committee on Appropriations, *Report to Accompany H.R. 9221 (DoD Appropriation Bill, 1966)*, Report No. 528, June 19, 1965, p. 49.

31 Eligibility of foreign entities to receive DoD basic research funding either directly or through a subaward varies based on the funding opportunity. DoD IOP interviewees indicated that there are multiple countries in which DoD cannot fund grants, and the countries that DoD cannot fund or visit are updated periodically.

32 Other approaches could also be appropriate for a mission-focused organization.

33 Documenting the funding sources for China's talent recruitment programs was outside the scope of the project.

34 NSFC, *2018 Annual Report*, English version, undated, p. 126.

35 See the Table 3.2 for examples of DoD use of the coordinated funding mechanism.

considered that archetype secondary for DOE. In contrast, NSF interviewees emphasized the effort that NSF staff put into creating joint funding calls with counterpart funders. NSF interviewees indicated that coordinated calls benefit the research community by eliminating the "double jeopardy" associated with the uncoordinated archetype in which the entire research partnership is at risk unless each nation independently decides to award funds. Accordingly, the team deemed coordinated funding a primary mechanism for NSF. Because NSF funding application materials are particularly clear about what information to provide about proposed foreign partners and activities, and NSF interviewees indicated that the organization supports a large volume of projects through the uncoordinated archetype, the team also considered uncoordinated funding a primary archetype for NSF. Within the set, NSF and DOE are relatively unique in supporting large-scale infrastructure collaborations. While supporting large-scale collaborative research infrastructure is part of the core missions of NSF and DOE, DoD, ANR, and NSFC have other responsibilities, and none appear to regularly support collaborative infrastructure projects.[36]

3.2.3. Incentives for Prospective Researchers

DoD interviewees indicated that DoD and entities in other countries use different approaches to try to position themselves as the partner of choice.[37] China offers generous amounts of funding but might require a researcher to move to or spend significant time in China. Many countries want to maintain the IP rights to the work they fund.[38] In contrast, DoD grants to international academics are comparatively small but come with favorable IP terms, permitting publication and allowing the researcher to retain the IP. (DoD gets a royalty-free license.) One DoD IOP indicated that collaborating with the United States and its scientists is often seen as prestigious, which can also help attract foreign PIs to work for U.S. organizations, such as DoD. Countering DoD's attractive characteristics, the connection to a defense organization deters some researchers (see Chapter Five).

Similar to DoD, ANR does not claim any IP rights from the projects it funds.[39] However, some programs have specific IP restrictions. For example, project proposals for the ASTRID and ASTRID Maturation programs that include foreign participants must include assurances

---

36 While NSFC has an agreement with CERN, China does not have CERN member or observer status suggesting that Chinese funders do not contribute to the facility itself. France is a CERN member, ANR's budget suggests that French contributions to collaborative infrastructure projects come from other organizations. ANR, *Soutenir la recherche dans sa diversité—Funding Research in All Its Diversity: Rapport d'Activité 2019—Annual Report 2019*, undated; CERN, "China," webpage, February 25, 2020; CERN, "Our Member States," webpage, undated.

37 One to two interviewees provided each observation.

38 Although basic research is normally nonproprietary and published openly, IP can nonetheless be a byproduct.

39 ANR, *Reglement Financier Relatif aux Modalités d'Attribution des Aides de l'ANR* [*Financial Regulations on the Terms of Allocation Aid from the ANR*], July 2019.

that the results from the French partners will be available for commercialization for defense purposes.[40]

The DOE allows IP ownership rights to be negotiated as part of international partnership agreements. For instance, national laboratories are often hired as technical performers where the IP rights are negotiated as part of the contract terms.[41]

3.2.4. Organizational Engagements with Researchers Abroad

The three U.S. organizations studied each use different approaches to engage with researchers outside of the United States, with DoD putting the most emphasis on identifying and forging relationships with individual researchers. For DoD, AFOSR/IO, ONR Global, and CCDC all station IOPs, for whom basic research engagement is a core part of their responsibilities, around the world. It is not uncommon for IOPs to be in contact with a promising foreign researcher for several years and to iterate through multiple ideas before making an award, which enables DoD to scope activities to address scientific and strategic goals. One DoD interviewee indicated that even though IOPs rotate every few years, one of the benefits of stationing personnel in-country is that it enables DoD to maintain an enduring presence and to become a trusted partner. A DOE interviewee indicated that while DOE also stations some employees abroad, they support all DOE activities, not just research. In contrast, in 2018, NSF closed its overseas offices, indicating that changes in the world had reduced need to have a physical presence in a location.[42] Instead, NSF deploys experts on short international visits through the MULTIPLIER (MULTIPlying Impact Leveraging International Expertise in Research Missions) program to build networks and explore opportunities for science and engineering collaboration.[43] An NSF interviewee indicated that NSF made the change to gain flexibility; rather than using a small number of dedicated staff in overseas offices, sending teams on focused visits allows NSF to draw on the expertise of a larger number of people using more of a bottom-up approach.

Additionally, both NSF and DOE engage with counterpart government organizations and the scientific communities about proposed and ongoing large-scale infrastructure. While DoD IOPs do engage with foreign government personnel on a country-by-country basis, it is often less of a priority than are direct engagements with researchers.

---

[40] ANR, *Accompagnement Spécifique des Travaux de Recherches et d'Innovation Défense: ASTRID* [*Specific Support for Defense Research and Innovation Work: ASTRID*], 2020, p. 13.

[41] DOE, *Bilateral DOE Laboratory Utilization Agreement*, May 26, 2010.

[42] NSF, "NSF Announces Changes to Overseas Offices," press release 18-003, February 21, 2018.

[43] NSF, OISE, "NSF MULTIPLIER—MULTIPlying Impact Leveraging International Expertise in Research Missions," webpage, undated.

CHAPTER FOUR

Scientific and Strategic Considerations for International Basic Research Collaboration

This chapter discusses factors that researchers and organizations often weigh when deciding whether to accept a collaboration opportunity. The chapter first discusses two categories of collaboration considerations: scientific and strategic.[1] Then, it describes specific considerations in each category along with their relevance to DoD. Finally, it presents a range of metrics that might be used to evaluate the considerations and identify collaboration opportunities.

A key factor in the attractiveness of any collaboration is how it would fit into a researcher or an organization's overall research needs. The *2020 Department of Defense International Science and Technology Engagement Strategy*, for example, emphasizes that international collaboration opportunities should be considered in the context of DoD's S&T needs, current efforts, and the associated gaps.[2] However, to focus on the considerations that are unique to international collaboration, the chapter does not examine how DoD or other organizations determine their overall S&T needs and priorities or how they evaluate international opportunities in the context of their research needs and gaps. The *2020 Department of Defense International Science and Technology Engagement Strategy* provides an illustrative example of how an organization might use a portfolio approach to combine information about its S&T priorities and gaps, its regional priorities, foreign S&T strengths, and foreign willingness and ability to partner.[3]

---

[1] The project did not examine the strategic and scientific considerations that influence overall organizational decisions about the research fields in which to invest (independent of where the research is performed).

[2] DoD, *Department of Defense International Science and Technology Engagement Strategy: A Unified Approach to Strengthen Alliances and Attract New Partners*, October 2020.

[3] DoD, *Department of Defense International Science and Technology Engagement Strategy: A Unified Approach to Strengthen Alliances and Attract New Partners*, October 2020.

4.1. Overview of Collaboration Considerations

Researchers and governments support IBRC for many reasons. For example, on the researcher level, international research collaborations can increase researcher productivity,[4, 5] help researchers to gain new skills and knowledge,[6] and provide opportunities for career advancement that might not be available nationally.[7] The benefits from collaboration to individual researchers aggregate over time, resulting in benefits to the national research enterprise.[8]

Government motivations to participate in international research collaborations include knowledge creation;[9] capacity-building for their own researchers,[10] as well as partners and allies;[11] an increased awareness of the research landscape;[12] and stronger relationships with partners and allies.[13] These outcomes can, in turn, support broader strategic goals of solving

4 Jonathan Adams, Karen Gurney, and Stuart Marshall, "Patterns of International Collaboration for the UK and Leading Partners (Summary Report)," *Evidence*, 2007.

5 Ram C. Acharya and Wolfgang Keller, *Technology Transfer Through Imports*, NBER Working Paper No. 13086, May 2007.

6 ERA Expert Group, *Opening to the World: International Cooperation in Science and Technology*, European Commission, EUR 23325 EN, 2008.

7 Jakob Edler, John Rigby, and Barbara Jones, *Review of Human Frontier Science Program Final Report*, Human Frontiers Science Program, 2010, p. 147.

8 Australian Academy of the Humanities, Department of Industry and Science, *Measuring the Value of International Research Collaboration*, May 2015.

9 Patries Boekholt, Jakob Edler, Paul Cunningham, and Kieron Flanagan, *Drivers of International Collaboration in Research*, Technopolis Group and the University of Manchester, April 1, 2009; and Luke Georghiou, "Global Cooperation in Research," *Research Policy*, Vol. 27, No. 6, September 1998.

10 Luke Georghiou, "Global Cooperation in Research," *Research Policy*, Vol. 27, No. 6, September 1998.

11 Caroline S. Wagner, Irene T. Brahmakulam, Brian A. Jackson, Anny Wong, and Tatsuro Yoda, *Science & Technology Collaboration: Building Capacity in Developing Countries?* RAND Corporation, MR-1357.0-WB, 2001.

12 Barry Bozeman and Jan Youtie, *The Strength in Numbers: The New Science of Team Science*, Princeton University Press, 2017.

13 Royal Society and American Society for the Advancement of Science, "New Frontiers in Science Diplomacy: Navigating the Changing Balance of Power," January 2010; Luke Georghiou, "Global Cooperation in Research," *Research Policy*, Vol. 27, No. 6, September 1998; National Academy of Sciences, National Academy of Engineering, and Institute of Medicine, *Examining Core Elements of International Research Collaboration: Summary of a Workshop*, National Academies Press, 2011.

grand challenges,[14] stimulating economic growth and development,[15] increasing security,[16] supporting innovation,[17] and promoting human development.[18]

Multiple authors have observed that international research collaborations generate benefits through both their direct scientific outputs and more indirect outcomes. For example, Wagner, 2006, found that international collaborations are attractive to governments for three reasons: (1) "to promote knowledge creation," (2) "to meet specific public missions," and (3) "to meet larger policy goals."[19]

Boekholt et al., 2009, argued that the drivers of international research collaborations can be grouped into a narrow paradigm and a broad paradigm.[20] The narrow paradigm focuses on using international research collaboration to improve the quality and quantity of scientific advances. The broad paradigm encompasses nonscientific objectives such as increasing national competitiveness, improving the scientific capabilities of developing countries, addressing societal challenges, and creating and strengthening diplomatic relationships.[21]

Table 4.1 shows some of the primary IBRC considerations for DoD identified by interviewees and in the literature and indicates whether each is mainly scientific or strategic. *Scientific considerations* are predominantly concerned with advancing science and are often connected to the benefits a collaboration might provide. *Strategic considerations* are primarily nonscientific and include factors like leadership desires to strengthen ties with another country or avoid strengthening a competitor's military capabilities. While some strategic considerations can be connected to benefits, such as supporting economic growth, strategic considerations more commonly characterize the risks of a collaboration. For example, working with a potentially untrustworthy partner in a scientific field where breakthroughs could

---

14 National Academy of Sciences, National Academy of Engineering, and Institute of Medicine, *Examining Core Elements of International Research Collaboration: Summary of a Workshop*, National Academies Press, 2011.

15 Australian Academy of the Humanities, Department of Industry and Science, *Measuring the Value of International Research Collaboration*, May 2015.

16 Patries Boekholt, Jakob Edler, Paul Cunningham, and Kieron Flanagan, *Drivers of International Collaboration in Research*, Technopolis Group and the University of Manchester, April 1, 2009.

17 L. Fleming, H. Greene, G. Li, M. Marx, and D. Yao, "Government-Funded Research Increasingly Fuels Innovation," *Science*, Vol. 364, No. 6446, June 21, 2019.

18 Caroline S. Wagner, Irene T. Brahmakulam, Brian A. Jackson, Anny Wong, and Tatsuro Yoda, *Science & Technology Collaboration: Building Capacity in Developing Countries?* RAND Corporation, MR-1357.0-WB, 2001.

19 Caroline S. Wagner, "International Collaboration in Science and Technology: Promises and Pitfalls," in Louk Box and Rutger Engelhard, eds., *Science and Technology Policy for Development, Dialogues at the Interface*, Anthem Press, 2006, p. 167.

20 Patries Boekholt, Jakob Edler, Paul Cunningham, and Kieron Flanagan, *Drivers of International Collaboration in Research*, Technopolis Group and the University of Manchester, April 1, 2009.

21 Patries Boekholt, Jakob Edler, Paul Cunningham, and Kieron Flanagan, *Drivers of International Collaboration in Research*, Technopolis Group and the University of Manchester, April 1, 2009.

TABLE 4.1
Considerations for International Basic Research Collaboration

Consideration	Scientific or Strategic?
Access to unique and diverse resources	Scientific
Ability to combine resources and share the burden	Both
Build and strengthen relationships	Both
Trust in a partner	Strategic
Economic impacts	Strategic
Defense impacts	Strategic
Maturity of a scientific field	Strategic
Speed of transition within a scientific field	Strategic
Amount of collaboration-related overhead	Strategic

rapidly be applied to create a military advantage is risky. As discussed later, considerations are frequently interconnected.

Multiple considerations exist beyond those listed in Table 4.1. For example, for some fields, such as ocean acoustics, DoD is one of many funders seeking to attract the top researchers. Ocean acoustics is of interest to DoD for its relation to naval sensing and communications, but the bulk of the research done in the field is for civilian applications and marine biology. In such fields, the low proportion of DoD funding might make it difficult for DoD to influence the research agenda. In those cases, partnering with other funders—either domestic or foreign—might help DoD increase the focus on its needs.

Additionally, collaborations norms, such as the typical number of researchers and institutions on a team, vary among fields. Well-targeted engagements in individualistic fields, where small teams or single investigators can make breakthroughs, might help reduce potential technological surprise. In highly collaborative fields, engaging with key nodes can provide visibility on large networks.[22]

Each collaboration is unique, and different considerations can dominate decisionmaking processes about whether to proceed with individual opportunities. Both bottom-up and top-down forces affect collaboration networks. For example, government funders establish policies that prohibit collaborations with some partners in some fields while issuing joint calls for proposals with some countries in other fields. Meanwhile, individual researchers independently identify opportunities and decide whether to pursue them.

---

[22] Ronald S. Burt, "Structural Holes and Good Ideas," *American Journal of Sociology*, Vol. 110, No. 2, September 2004.

To examine how the considerations for IBRC differ among research fields of interest to DoD, the project team assembled a diverse set of candidate research fields from two lists (neither list focused on DoD IBRC priorities and most included work at a range of TRLs):

1. the Under Secretary of Defense for Research and Engineering (USD[R&E]) Modernization Priorities List, which contains near- and medium-term priorities and is broader than basic research[23]
2. the FY20 Defense Established Program to Stimulate Competitive Research (DEPSCoR) topic areas, which is a basic research program to increase capability at U.S. institutes of higher education.[24]

The team combined the lists and excluded topics deemed too narrow or broad or outside of traditional DoD strengths. The team then added mathematics and synthetic biology for sensors and materials to the list at the sponsor's suggestion. While all topics in the set are DoD priorities, the set was not constructed to represent either the top basic research priorities or the top priorities for IBRC. Instead, the set was intended to provide examples of the variety of IBRC considerations that arise in scientific topics of interest to DoD.

Collaboration opportunities within a scientific field are often subject to similar forces making the resulting collaboration patterns field-dependent (see Figure 4.1). Magnetospheric plasma, for example, relies on expensive, shared infrastructure. This need for satellite and telescope infrastructure coupled with the low risks associated with immediate economic or defense impacts has led to U.S. researchers working with non-U.S. researchers on the majority of their publications. In contrast, the field of energetic materials is subject to some export control regulations as well as concerns about near-term military applications; thus, few energetic materials papers with a U.S. author have a non-U.S. co-author.

4.2. Considerations

This section describes each of the collaboration considerations listed in Table 4.1 along with the associated implications for DoD. Appendix C elaborates by providing an overview of the considerations for two fields: microelectronics and optoelectronics.

4.2.1. Access to Unique and Diverse Resources

Multiple interviewees observed that the majority of the world's research is being conducted outside of the United States, with one-of-a-kind facilities and materials as well as the best and brightest scientists located throughout the world. International cooperation can improve

[23] MITRE, "USD(R&E) Top 10 Technology Focus Areas," webpage, undated.

[24] AFOSR, *Funding Opportunity Announcement: FY20 Defense Established Program to Stimulate Competitive Research (DEPSCoR)*, undated.

FIGURE 4.1

Percentage of Papers in Example Fields with U.S. and Non-U.S. Authors

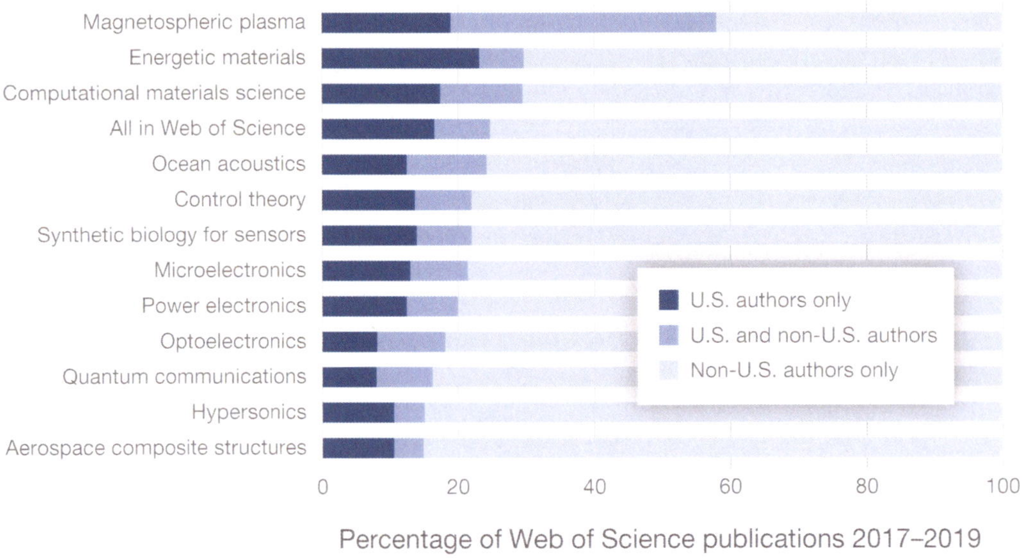

NOTES: The team used Web of Science's country code to categorize publications as having U.S. authors, non-U.S. authors, or both. See the preceding pages for detail on how the project team selected the scientific fields. The project team used the following queries for the Web of Science searches: aerospace composite structures: TS=((“airframe*” or “aerospace” or “aircraft*” or “airplane*”) and (“composite*” and “structur*”)); computational materials science: TS=(computational) and WC=(“materials science”); control theory: TS=(“control theory”); energetic materials: TS=(“energetic material*”); hypersonics: TS=(hypersonic*); magnetospheric plasma: TS=(“magnetospher*” and “plasma*”); microelectronics: TS=(“microelectronic*”); ocean acoustics: TS=(((ocean or underwater or sea) and acoustic*) or hydroacoustic); optoelectronics: TS=(optoelectronic*); power electronics: TS=((“power electronic*” or “Gallium Nitride” or “GaN” or “wide-bandgap semiconductor*”) not (asthma or adversar*)); quantum communications: TS=(“quantum communication*” or “quantum key distribution” or “quantum cryptography” or “quantum encryption”); and synthetic biology for sensors: TS=(((“*synthetic*” OR “*engineer*”) AND (“biosensor*” or “bio-sensor*”)) OR (“synthetic biology” and *sensor*)).

research quality by giving researchers access to resources not available in their own countries.[25, 26, 27] In explaining the increasing need for collaboration, HHS notes that research now calls for not only resources but *specific* resources to allow investigations in “increasingly novel, complex and convoluted areas.”[28] The resources needed to achieve a specific project's

---

25 CREST Working Group, *Internationalisation of R&D—Facing the Challenge of Globalisation: Approaches to a Proactive International Policy in S&T*, November 27, 2007.

26 Patries Boekholt, Jakob Edler, Paul Cunningham, and Kieron Flanagan, *Drivers of International Collaboration in Research*, Technopolis Group and the University of Manchester, April 1, 2009.

27 Royal Society and American Society for the Advancement of Science, “New Frontiers in Science Diplomacy: Navigating the Changing Balance of Power,” January 2010; Luke Georghiou, “Global Cooperation in Research,” *Research Policy*, Vol. 27, No. 6, September 1998; National Academy of Sciences, National Academy of Engineering, and Institute of Medicine, *Examining Core Elements of International Research Collaboration: Summary of a Workshop*, National Academies Press, 2011.

28 HHS, Office of Research Integrity, “Need for Collaboration,” webpage, undated.

goals may not be available nationally, so collaborating with international partners is a way to gain access to key resources.

A variety of resources are available through IBRC:

- **Expertise:** Access to resources includes the ability to work and interact with research experts abroad or with foreign sources of knowledge.[29, 30, 31, 32, 33] For example, the Bilateral Academic Research Initiative enabled the U.S. DoD and the UK Ministry of Defence to access artificial intelligence (AI) experts in both countries.[34]
- **Facilities:** For example, the Hypersonic International Flight Research Experimentation Program, which included AFRL, the Australian Defence Science and Technology Group, and industry and academic partners, used the Woomera test range, a unique test facility located in Australia.[35, 36] Similarly, DOE has supported collaborations that enabled U.S. researchers to access unique tokamak facilities in the EU, the Republic of Korea, and China.[37, 38]
- **Tools:** A DoD interviewee indicated that Europe is often more willing to support maintenance and ongoing development of resources for the scientific community than the United States. As an example, the interviewee pointed to decades-long European support for the laboratories that develop and maintain the codes for the computational materials science field, indicating that a previous U.S. effort to duplicate the resources was prohibitively expensive and unsustainable.

---

29 CREST Working Group, *Internationalisation of R&D Facing the Challenge of Globalisation: Approaches to a Proactive International Policy in S&T*, November 27, 2007.

30 Jonathan Adams, Karen Gurney, and Stuart Marshall, "Patterns of International Collaboration for the UK and Leading Partners (Summary Report)," *Evidence*, 2007.

31 Patries Boekholt, Jakob Edler, Paul Cunningham, and Kieron Flanagan, *Drivers of International Collaboration in Research*, Technopolis Group and the University of Manchester, April 1, 2009.

32 The Royal Society and American Society for the Advancement of Science, "New Frontiers in Science Diplomacy: Navigating the Changing Balance of Power," January 2010; Luke Georghiou, "Global Cooperation in Research," *Research Policy*, Vol. 27, No. 6, September 1998; National Academy of Sciences, National Academy of Engineering, and Institute of Medicine, *Examining Core Elements of International Research Collaboration: Summary of a Workshop*, National Academies Press, 2011.

33 Australian Academy of the Humanities, Department of Industry and Science, *Measuring the Value of International Research Collaboration*, May 2015.

34 DoD, "DoD Announces BARI Award for US-UK Collaboration on Human-Machine Teaming," press release, September 25, 2018.

35 Boeing, "Speeding Up Hypersonic Innovation with HIFiRE 4," webpage, August 10, 2017.

36 Defence Science and Technology Group, Australian Government Department of Defence, "HIFIRE PROGRAM," webpage, undated.

37 World Nuclear News, "DOE Announces Funds for Overseas Fusion Research," webpage, March 11, 2019.

38 Princeton Plasma Physics Laboratory, "U.S.-China Collaboration Makes Excellent Start in Optimizing Lithium to Control Fusion Plasmas," press release, June 22, 2017.

4.2.2. Ability to Combine Resources and Share the Burden to Advance Scientific or Strategic Goals

Research is often resource intensive. Collaborations that combine resources can help partners lower research costs and avoid unintended duplication on both small and large scales. Synchronizing research priorities and standardizing data and methods can promote resource sharing success.

4.2.2.1. Address Global Challenges

Global challenges that are too expansive for one country to tackle necessitate international collaboration.[39] The *Summary of the 2018 National Defense Strategy of the United States of America* notes the contribution that pooling resources makes towards reducing the security burden.[40] Similarly, DOE seeks to share the burden of addressing challenges related to "nonproliferation, security, safety, and the environment" with partners and allies.[41]

Increasing interest in focusing research agendas around grand societal challenges has been accompanied by an increase in international research collaborations focused on burden-sharing.[42] ANR notes that pooling resources allows it and international partners to "provide joint responses to global challenges that cannot be addressed solely at national level."[43] Common problems that IBRC has worked to address include climate change and, especially during the time this report was being written, pandemics.

4.2.2.2. Build and Maintain Large-Scale Instruments, Infrastructure, and Resources

IBRC allows expensive and specialized facilities and resources, such as particle accelerators and large telescopes, to be shared. One prominent example is CERN, which operates a particle accelerator near Geneva, Switzerland, that is capable of producing the highest particle energies in the world. NSF has contributed to the construction of detectors at CERN and has entered into agreements that allow U.S. scientists to operate experiments in the European

---

39 John Kania and Mark Kramer, "Collective Impact," *Stanford Social Innovation Review*, Winter 2011.

40 Jim Mattis, *Summary of the 2018 National Defense Strategy of the United States of America: Sharpening the American Military's Competitive Edge*, U.S. Department of Defense, undated, p. 8.

41 DOE, Office of Nuclear Energy, "International Nuclear Energy Policy and Cooperation," webpage, undated.

42 Aurelia Stamm and Andreas Figueroa, "Effective International Science, Technology and Innovation Collaboration: From Lessons Learned to Policy Change," in OECD, *Meeting Global Challenges Through Better Governance: International Co-operation in Science, Technology and Innovation*, OECD Publishing, 2012.

43 ANR, "Transnational Cooperation Strategy," webpage, undated.

facility while allowing European scientists to work at U.S. detectors.[44, 45] Similarly, DOE has provided researchers and in-kind support to CERN.[46]

Another example is the Extreme Light Infrastructure (ELI), a set of three high-energy laser facilities located in the Czech Republic, Hungary, and Romania that as of 2021 was in its initial operations phase.[47] ELI was funded through a combination of contributions from the countries hosting the facilities and EU contributions, including money from infrastructure development funds that had previously been used only for nonscientific infrastructure (e.g., roads, power distribution systems).[48] While the ELI project has experienced some disputes, the resulting facility is expected to advance multiple scientific fields, including biomedicine, astrophysics, and materials science.[49]

4.2.2.3. Pool Project Resources

IBRC can also allow countries to pool resources on a smaller scale. For example, through the International Cooperative Engagement Program for Polar Research (ICE-PPR), ONR works with partners from Canada, Denmark, Finland, Iceland, New Zealand, Norway, and Sweden to study the Arctic.[50] In an ICE-PPR project to deploy research buoys to collect data on ocean activity in the Arctic, the Royal Danish Air Force provided aircraft, Canada provided buoys, and ONR provided logistical support, illustrating the range of resources that partners can provide.[51] Researchers also frequently combine resources through bottom-up, investigator-initiated collaborations.

Whether a collaboration is large or small, reciprocity is typically expected. NSFC's public materials communicate that international collaborators are to share benefits equally.[52] Similarly, balance in financial and scientific contributions as well as IP rights is a fundamental characteristic ANR expects to be present in international collaboration proposals.[53]

---

44 NSF, *NSF FY 2021 Budget Request to Congress: High Luminosity-Large Hadron Collider Upgrade (HL-LHC)*, undated.

45 NSF, "The Excitement Behind a Renewed Agreement," webpage, May 7, 2015.

46 CERN, "US Becomes Observer at CERN," press release, December 19, 1997.

47 ELI, "The ELI Project," webpage, undated.

48 ELI, "The ELI Project," webpage, undated.

49 Alison Abbott, "World's Largest Laser Lab Rocked by Slew of Disputes," *Nature*, No. 569, May 23, 2019 (corrected May 28, 2019).

50 Warren Duffie, Jr., "Arctic Ambition: ONR Joins International Mission to Monitor Conditions in Arctic Ocean," Office of Naval Research Public Affairs, September 14, 2017.

51 Warren Duffie, Jr., "Arctic Ambition: ONR Joins International Mission to Monitor Conditions in Arctic Ocean," Office of Naval Research Public Affairs, September 14, 2017.

52 NSFC, *National Natural Science Fund Guide to Programs 2019*, English version, undated, p. 193.

53 ANR, 2021 *Work Programme: Generic Call for Proposals: 2021 AAPG Guide: Submission, Evaluation, Selection and Funding Guidelines*, English version, v.1.0, October 2, 2020, p. 14.

4.2.2.4. Leverage Other Countries' Prior Investments to Maximize the Value of Research Funds

Multiple DoD interviewees indicated that DoD also looks for opportunities to share the burden by leveraging other countries' prior investments.[54] A foreign PI receiving funding, including defense funding, from a friendly or allied government is usually considered as a positive as it confirms interest in the work and gives DoD the opportunity to leverage investments from trusted sources. DoD can sometimes make a large impact with a small grant to an international PI that leverages large amounts of current or recent investment from other organizations, such as a European Research Council grant that was used to create a lot of data and tools. DoD grants to foreign academic researchers can be an economical way to test an idea before making investments in the service laboratories. The reduced research and overhead costs in some countries compared with the United States can further increase the value of such grants to DoD. Similarly, providing grants to foreign researchers who can access their countries' facilities at low or no cost can be more economical than paying the fees for a U.S. researcher to access the facility or creating a similar facility in the United States. For example, LaserLab Europe provides preferred access to EU citizens.[55]

4.2.3. Build and Strengthen Relationships

Relationship and network building in international research collaborations happens on multiple levels—researcher to researcher, institution to institution, and government to government—and is both foundational to effective collaboration and a benefit of collaboration.[56]

4.2.3.1. Contribution of Relationships to Scientific Strength

A robust network of global relationships is an important asset whose value extends beyond the benefits of individual partnerships. The United States' position in the global research network matters because central actors in a knowledge network have superior access to knowledge.[57] Visibility and awareness can reduce the likelihood of technological surprise. Innovation in many fields is the product of a novel recombination of ideas, which is facilitated by collaboration.[58] The networks of collaborators are, themselves, important components of innova-

---

[54] One to two interviewees provided each illustration.

[55] Laserlab-Europe, "Using Our Facilities," undated.

[56] Caroline S. Wagner, "International Collaboration in Science and Technology: Promises and Pitfalls," in Louk Box and Rutger Engelhard, eds., *Science and Technology Policy for Development, Dialogues at the Interface*, Anthem Press, 2006.

[57] Ronald S. Burt, *Structural Holes: The Social Structure of Competition*, Harvard University Press, 1992.

[58] Sari Pekkala Kerr and William R. Kerr, "Global Collaborative Patents," *The Economic Journal*, Vol. 128, No. 612, 2018.

tion that bring diverse groups of specialists into contact.[59] Networks can provide awareness of global scientific activities and developments, help researchers identify new opportunities, and transform the way domestic researchers conduct studies.[60,61] International collaborations also provide opportunities to train U.S. scientists and engineers in the international research environment, which is important for sustaining relationships in an increasingly international research and education environment.[62]

As described in the *AFOSR International Office FY19 Annual Report*, when the Air Force invests in PIs, the Air Force also gains access to their networks, which are the products of much larger investments.[63] As former award recipients advance in their careers, the size and value of those network connections increase.[64] Because approximately 80 percent of researchers are based outside of the United States, international activities can significantly increase the size of the networks to which DoD researchers belong.[65] Having established networks has helped the Air Force respond to crises, including the global COVID-19 pandemic.[66]

The vision painted in the *2020 Department of Defense International Science and Technology Engagement Strategy* includes "[t]hriving international networks."[67] Multiple DoD IOP interviewees indicated that one of their objectives is to build and maintain a network of researchers. The outreach that IOPs conduct connects DoD with researchers who might not otherwise engage with the organization.

Encouraging partnerships and multidisciplinary collaborations is fundamental to ANR's mission, and per ANR's Transnational Cooperation Strategy, the organization seeks to "help researchers put together lasting scientific networks" and to "[r]aise the profile and strengthen the influence of French research on the international stage."[68] Similarly, NSFC views confer-

---

59 Walter W. Powell, Kenneth K. Koput, and Laurel Smith-Doerr, "Interorganizational Collaboration and the Locus of Innovation: Networks of Learning in Biotechnology," *Administrative Science Quarterly*, Vol. 41, No. 1, 1996.

60 NRC, *S&T Strategies of Six Countries: Implications for the United States*, National Academies Press, 2010.

61 Emilia Primeri, Emanuela Reale, Benedetto Lepori, Philippe Laredo, Maria Nedeva, and Duncan Thomas, "Measuring the Opening of National R&D Programs: What Indicators for What Purposes?" *Research Evaluation*, Vol. 23, No. 4, October 2014.

62 NSF, OISE, "International Activities at NSF," webpage, undated.

63 AFRL, AFOSR, *AFOSR International Office FY19 Annual Report*, undated, p. 4.

64 AFRL, AFOSR, *AFOSR International Office FY19 Annual Report*, undated, pp. 4–5.

65 AFRL, AFOSR, *AFOSR International Office FY19 Annual Report*, undated, p. 5.

66 AFRL, "#BasicResearch Chatter—An Opportunity to Chat About the AFRL Engineer and Scientist Exchange Program (ESEP)," *Afrl.dod.live.mil*, December 14, 2020.

67 DoD, *Department of Defense International Science and Technology Engagement Strategy: A Unified Approach to Strengthen Alliances and Attract New Partners*, October 2020, p. 6.

68 ANR, "Transnational Cooperation Strategy," webpage, undated.

ences and short-term personnel exchanges as a way to form networks that could lead to more substantial future collaborations as well as increase the influence of Chinese researchers.[69]

4.2.3.2. Using Relationships from International Basic Research Collaborations to Advance Strategic Goals

In what is often called *science for diplomacy*, scientific collaborations that work to solve problems of common interest can be used for diplomatic purposes in addition to knowledge creation.[70] The soft power of science diplomacy can help countries maintain channels of communication when relations are deteriorating or just beginning, can open doors for future cooperation within science or other areas of diplomatic interest, and can improve the countries' overall relationship due to spillover effects.[71, 72, 73]

For example, beginning in the 1950s, and continuing throughout the Cold War, the United States and the Union of Soviet Socialist Republics (USSR) collaborated on research on Antarctica, outer space, and oceanography.[74] The resulting people-to-people ties have been credited with building trust and understanding that contributed to improvements in the overall relationship between the countries.[75, 76] Additionally, Russian post–Cold War involvement with the International Space Station has served as a foreign policy tool, in addition to allowing the original partners to further share the burden and to benefit from Russian space expertise.[77] Similarly, the Synchrotron-light for Experimental Science and Applications in the Middle East (SESAME) has both scientific objectives and the goal to create connections between different societies in the Middle East and its neighbors with the hope that these connections can improve understanding and tolerance between nations.[78]

---

69 NSFC, *2018 Annual Report*, English version, undated, p. 41.

70 Royal Society and American Society for the Advancement of Science, "New Frontiers in Science Diplomacy: Navigating the Changing Balance of Power," January 2010.

71 Royal Society and American Society for the Advancement of Science, "New Frontiers in Science Diplomacy: Navigating the Changing Balance of Power," January 2010.

72 ERA Expert Group, *Opening to the World: International Cooperation in Science and Technology*, European Commission, EUR 23325 EN, 2008.

73 Australian Academy of the Humanities, Department of Industry and Science, *Measuring the Value of International Research Collaboration*, May 2015.

74 Paul Arthur Berkman, "Could Science Diplomacy Be the Key to Stabilizing International Relations?" *Scientific American*, June 12, 2018.

75 Paul Arthur Berkman, "Could Science Diplomacy Be the Key to Stabilizing International Relations?" *Scientific American*, June 12, 2018.

76 Olga Krasnyak, "How U.S.-Soviet Scientific and Technical Exchanges Helped End the Cold War," *American Diplomacy*, November 2019.

77 Julie Payette, "Research and Diplomacy 350 Kilometers Above the Earth," *Science & Diplomacy*, December 10, 2012.

78 SESAME, "What Is SESAME?" webpage, undated.

Multiple DoD IOP interviewees observed that basic research is often the easiest place to start engaging another country and that relationships can provide opportunities to influence areas beyond basic research.[79] Being stationed abroad creates opportunities for DoD IOPs to be drawn into other roles, such as helping other U.S. scientific agencies (e.g., NSF and DOE) develop their own partnerships and accepting invitations to attend foreign government meetings and provide advice. Working with foreign academic communities can help the United States build bonds and trust with the researchers' countries; this trust can then lead to future collaboration on defense applications. In the longer term, those collaborations could result in increased military interoperability, which the *Summary of the 2018 National Defense Strategy of the United States of America* identifies as one of the ways to achieve a capable network of partners and allies.[80]

Multiple DoD interviewees discussed the importance of DoD (and the United States) being the partner of choice.[81] Top researchers have multiple collaboration opportunities available to them. By engaging internationally and working to be the partner of choice, the United States can reduce the chances of other countries, such as strategic competitors, becoming the partner of choice. DoD IOP interviewees considered such efforts to be particularly critical in parts of the world, such as South America, where the United States is a comparatively recent partner and collaboration patterns have not yet been set. Interviewees feared that if DoD did not engage, it would leave an opening for another country to become a preferred partner and reap the associated benefits.

While both NSFC and ANR engage worldwide, they have geographic focuses coupled to political goals. NSFC prioritizes collaborations that support China's Belt and Road Initiative and "one country, two systems" policies.[82] ANR, consistent with France's membership in the EU and its support for the European Research Area (ERA), focuses its international cooperation energy on Europe. Multiple ANR publications use the phrase "European and international," illustrating the partitioning of the set of potential partners into Europe and everyone else. Within Europe and in accordance with the provisions of the Franco-German Treaty of Aachen that calls for France and Germany to establish joint research programs and funding mechanisms, ANR prioritizes French-German collaborations.[83, 84]

Playing a prominent role in scientific networks can also enhance national prestige and influence, including the ability to shape scientific norms. For example, China has been using

---

[79] One or two interviewees provided each illustration.

[80] Jim Mattis, *Summary of the 2018 National Defense Strategy of the United States of America: Sharpening the American Military's Competitive Edge*, U.S. Department of Defense, undated.

[81] One or three interviewees provided each illustration.

[82] NSFC, *National Natural Science Fund Guide to Programs 2019*, English version, undated, p. 199.

[83] ANR, *Soutenir la recherche dans sa diversité—Funding Research in All Its Diversity: Rapport d'Activité 2019—Annual Report 2019*, undated, p. 35.

[84] Federal Republic of Germany and the French Republic, Treaty Between the Federal Republic of Germany and the French Republic on Franco-German Cooperation and Integration, January 22, 2019.

its increasing global scientific prominence to influence international standards.[85] Interacting globally on research on emerging technologies, such as AI, or participating in scientific meetings could help China further expand its impact.[86] Historic research leadership has helped the United States to propagate its approach to ethics, norms, and values, inculcating these both in U.S. researchers and those who train or collaborate with them. DOE international research collaborations help the United States increase its credibility and influence in activities associated with nuclear technologies and nonproliferation.[87] Additionally, an interviewee indicated that collaborating on large international projects gives the DOE a seat at the table, which allows it to advocate for the U.S. scientific community.

4.2.4. Trust in a Partner

How much trust—if any—to place in a partner is one of the key decisions in any collaboration. According to a 2019 JASON report, "[b]roadly considered, participants in collaborations should practice transparency, reciprocity, and adherence to norms of research integrity, and should expect the same from other participants."[88] As an example, a project interviewee provided anecdotal support about asymmetric access for international collaborators, with Chinese visitors able to see people, material, and facilities when visiting the United States, but U.S. visitors discovering that they were unable to go where they expected while visiting their collaborators in China.

Violations of research integrity, including hidden diversions of resources and misrepresentation of associations (see Chapter One), are a betrayal of trust. A 2019 DoD letter to academic researchers urges balance, saying that "[o]pen international collaborations are important to DoD and the Nation, and we must also protect against those who seek to exploit this openness."[89] The letter refers to "theft of controlled information and unethical diversion of intellectual capital" as threats to research integrity, which is otherwise associated with "principles, such as free exchange of ideas, transparency, and collaboration across research communities."[90]

Trust matters on the individual, organizational, and country levels. When trust is low—either due to past betrayals or because a relationship is new—the importance of other considerations often increases. With a trusted ally, the added risk associated with collaborating

---

[85] U.S.-China Economic and Security Review Commission, *2019 Report to Congress of the U.S.-China Economic and Security Review Commission*, U.S. Government Publishing Office, November 2019.

[86] U.S.-China Economic and Security Review Commission, *2019 Report to Congress of the U.S.-China Economic and Security Review Commission*, U.S. Government Publishing Office, November 2019.

[87] DOE, Office of Nuclear Energy, "Bilateral Cooperation," webpage, undated.

[88] JASON, *Fundamental Research Security*, MITRE Corporation, December 2019, p. 28.

[89] Michael D. Griffin, *Letter to Academic Colleagues*, U.S. Department of Defense, October 10, 2019.

[90] Michael D. Griffin, *Letter to Academic Colleagues*, U.S. Department of Defense, October 10, 2019.

in fast-advancing fields or in fields with high economic and military implications might be acceptable; with a new partner, it likely would not be acceptable.

Country policies that place obligations on researchers that are inconsistent with U.S. research norms can challenge the trust needed for a collaboration. For example, a recent requirement that scientific data generated in China be reviewed by a new government-affiliated data center prior to its publication or release outside of the country[91] is a fundamental challenge to scientific openness and academic freedom. It complements China's "External Transfer of Intellectual Property Measures," which require security review where IP might be transferred or shared.[92] Another challenge comes from China's National Intelligence Law, which was enacted in 2017, which "obliges individuals, organizations, and institutions to assist Public Security and State Security officials in carrying out a wide array of 'intelligence' work."[93]

Part of this consideration is the ability of a partner to use research results in ways that are inconsistent with the wishes of other partners. Typically, one would only collaborate with a potential partner who has both the intent and capability to apply research results in a way that opposes the interests of other partners on low risk topics. For example, an NAE panel noted that the ability of adversaries to apply technologies to military applications impacts the appropriate level of concern regarding loss of information.[94]

Capabilities of adversarial nations, and the associated concerns, can change over time. In 1982, the aforementioned NAE panel observed that it is "current U.S. policy to help the People's Republic of China (PRC) advance its industrial technology. It is generally recognized that the capacity of the PRC to transfer such technologies to the military sector is limited."[95] Since then, China's scientific and military capabilities have grown considerably. Additionally, "military-civil fusion," the explicit leveraging of S&T for military aims, is now a Chinese national strategic priority.[96]

Consistent with military-civil fusion, most of China's civilian universities have some connection to China's defense industry. Departments of leading civilian universities "require

91 Dennis Normile, "China Asserts a Firm Grip on Research Data," *Science*, April 9, 2018; and Office of the State Council, People's Republic of China, "Notice of the General Office of the State Council on Issuing the Measures for the Administration of Scientific Data [国务院办公厅关于印发科学数据管理办法的通知]," April 2, 2018.

92 United Nations Conference on Trade and Development, "China—Introduces 'Measures for the Overseas Transfers of Intellectual Property Rights (Trial),'" webpage, March 29, 2018.

93 Murray Scot Tanner, "Beijing's New National Intelligence Law: from Defense to Offense," *Lawfare*, July 20, 2017.

94 National Academy of Engineering, *Scientific Communication and National Security*, National Academies Press, 1982.

95 National Academy of Engineering, *Scientific Communication and National Security*, National Academies Press, 1982.

96 U.S. Department of State, "Military-Civil Fusion and the People's Republic of China," fact sheet, May 28, 2020.

their researchers to hold advanced security credentials, receive funding from China's Ministry of State Security, or contribute directly to [People's Liberation Army]-led research projects."[97] In response to the publicity associated with such ties and USG research security concerns, U.S. universities seem to be exercising more caution in collaborations with China.[98]

Multiple DoD IOPs indicated that when deciding whether to make an award, they consider information that applicants provide about their background and past and current funding sources and consult with defense intelligence if anything appears suspicious.[99, 100] While past, present, or pending connections to a strategic competitor would not necessarily disqualify a potential foreign grantee, they would necessitate a more detailed security assessment that considers the sensitivity of the topic and whether the potential benefit is worth the risk. A particular concern is the presence of obligations that restrict the open collaboration that DoD is trying to facilitate.

4.2.5. Economic Impacts

Long-term economic growth is one reason why governments support basic research. NSF views the potential benefits of expanded partnerships as leveraging expertise and resources in pursuit of innovations; enhancing research, education, and workforce capacity; and improving translation from discovery to products and services that benefit society.[101] Additionally, international collaboration can help familiarize a government with another country's innovation landscape and the associated opportunities for future exports.[102, 103] Commercialization and technology maturation are central to the mission of DOE research, and partnerships are intended to provide economic benefits by developing new technologies and opening new export markets that will produce U.S.-based jobs or other economic advantages.[104] Likewise,

---

[97] Ryan Fedasiuk and Emily Weinstein, "Universities and the Chinese Defense Technology Workforce," *Center for Security and Emerging Technology*, December 2020, p. 17; Alex Joske, "The China Defence Universities Tracker," Australian Strategic Policy Institute, November 2019.

[98] See, for example, the statement in a Cornell University webpage advising faculty: Cornell University, "International Engagement for Faculty," webpage, undated.

[99] One to three interviewees provided each example.

[100] Interviewees did not elaborate on what types of information might appear to be suspicious.

[101] NSF, *Building the Future: Investing in Discovery and Innovation: NSF Strategic Plan for Fiscal Years (FY) 2018–2022*, NSF 18-45, February 2018.

[102] CREST Working Group, *Internationalisation of R&D—Facing the Challenge of Globalisation: Approaches to a Proactive International Policy in S&T*, November 27, 2007.

[103] Australian Academy of the Humanities, Department of Industry and Science, *Measuring the Value of International Research Collaboration*, May 2015.

[104] DOE, *Foreign Engagements with DOE National Laboratories*, DOE P 485.1A, December 13, 2019.

one reason DoD participates in international research collaborations is to maintain U.S. competitiveness in the global economy and militarily.[105]

However, as discussed in the following sections and Appendix C, economic impacts of some collaborations might also include eroding national strengths in science or reducing the benefits that the United States realizes from the eventual commercial or military applications of basic research results.

4.2.6. Defense Impacts

Even among entities that advocate for widespread openness in basic research, strong connections between a scientific field and "identifiable direct military applications" have been acknowledged as a reason to consider restricting IBRC, particularly in combination with other considerations, such as the potential for rapid transition from basic to applied research.[106] As a former OSTP director, John Holdren, explained to Congress, the potential for a large military or commercial advantage to the country that achieves a breakthrough can constrain international collaboration in basic research.[107] However, he also warned about the costs of being overly cautious and unnecessarily restricting IBRC.[108]

For example, from 1999 through 2014, the U.S. Munitions List of the U.S. International Traffic in Arms Regulations (ITAR) included all space satellites, including commercial ones.[109] Although ITAR typically does not apply to fundamental research at universities, much of the university work in the field included private companies and government laboratories that cannot use the ITAR exemption available to academic institutions.[110] The desire of potential foreign fundamental research collaboration partners to avoid overhead associated with determining what was necessary to comply with the U.S. ITAR and to obtain the associated licenses and technical assistance agreements, coupled with uncertainties and fear of criminal consequences, made the United States a less desirable partner and reduced its

---

[105] NRC, *Strategic Engagement in Global S&T: Opportunities for Defense Research*, National Academies Press, 2014.

[106] National Academy of Engineering, *Scientific Communication and National Security*, National Academies Press, 1982, p. 5.

[107] John P. Holdren, "Statement of the Honorable Dr. John P. Holdren, Director, Office of Science and Technology Policy, Executive Office of the President of the United States, to the Subcommittee on Oversight and Investigations, House Committee on Foreign Affairs," November 2, 2011.

[108] John P. Holdren, "Statement of the Honorable Dr. John P. Holdren, Director, Office of Science and Technology Policy, Executive Office of the President of the United States, to the Subcommittee on Oversight and Investigations, House Committee on Foreign Affairs," November 2, 2011.

[109] National Oceanic and Atmospheric Administration, Office of Space Commerce, "Revised Satellite Export Control Rules Published," webpage, May 13, 2014.

[110] NRC, *Space Science and the International Traffic in Arms Regulations: Summary of a Workshop*, National Academies Press, 2008.

ability to leverage foreign investments and access the associated diverse resources (including expertise).[111]

In some fields, such as information and communications technology (ICT), which includes semiconductors, a strong domestic industry is valuable not only for the economic benefits it provides but because it helps provide secure supply chains for DoD.[112] Concerns that the United States is losing ground to China in the development of advanced semiconductors have garnered national attention, and the U.S. Congress has debated funding initiatives to support continued American innovation and manufacturing in the field.[113] (See Section C.1.1 for information on the impact of government incentives on the selection of semiconductor manufacturing locations.) While funding of domestic science and manufacturing is the primary concern of many of these initiatives, there is also interest in using international collaboration to provide the United States access to needed resources. As part of the initiative to build U.S. semiconductor competitiveness, the United States has been working to develop strategic cooperation initiatives with Taiwan, including S&T collaborations.[114] Similarly, the Cyberspace Solarium Commission, which examined ICT, recommended supporting domestic projects as well as "multilateral research and development initiatives with partners and allies."[115]

The relationship between where foundational research is performed and where discoveries are commercialized is complicated. Even when manufacturing and R&D are tightly coupled, the extent to which R&D activities accompany manufacturing activities that relocate to a new country can vary by field (see Section C.1.1).[116]

4.2.7. Maturity of a Scientific Field

In mature scientific fields, the rate of change is often relatively predictable as researchers engage in problem-solving rather than seeking fundamental breakthroughs.[117] Conversely, newer fields of research often explore numerous lines of inquiry in parallel, and it is typi-

---

[111] NRC, *Space Science and the International Traffic in Arms Regulations: Summary of a Workshop*, National Academies Press, 2008.

[112] United States of America Cyberspace Solarium Commission, "Building a Trusted ICT Supply Chain: CSC White Paper #4," October 2020.

[113] David E. Sanger, Catie Edmondson, David McCabe, and Thomas Kaplan, "Senate Poised to Pass Huge Industrial Policy Bill to Counter China," *New York Times*, June 7, 2021.

[114] Lee Hsin-fang and William Hetherington, "Taiwan, US to Focus on Semiconductors," *Taipei Times*, November 22, 2020, p. 1.

[115] United States of America Cyberspace Solarium Commission, "Building a Trusted ICT Supply Chain: CSC White Paper #4," October 2020, p. 22.

[116] NRC, *Optics and Photonics: Essential Technologies for Our Nation*, National Academies Press, 2013, pp. 186–202.

[117] Thomas S. Kuhn, *The Structure of Scientific Revolutions*, University of Chicago Press, 1962.

cally difficult to predict which research paradigm will be most successful.[118] Partnering with research groups pursuing different elements within a field can reduce the potential of technological surprise by providing visibility when one research paradigm is nearing a potential breakthrough.

4.2.8. Speed of Transition Within a Scientific Field

"Fundamental Research," a 2010 DoD memorandum that continues to serve as the source for program management guidance, indicates that fundamental research generally should be managed so that it does not become subject to export controls, publication restrictions, or restrictions on the participation of foreign researchers.[119] In order to ensure that their grants remain in compliance, one interviewee reported that IOPs often consider as a red flag the potential for a topic to rapidly become applied.

The anticipated speed of commercialization can also interact with other considerations. In fast-moving fields, bureaucratic delays in starting a partnership can mean costly lost opportunities. Additionally, should IP-related issues arise, the wide variations in legal frameworks and the inefficiency of enforcement institutions in foreign countries can be exacerbated in rapidly moving industries.[120] An NAE report, *Scientific Communication and National Security*, found that the ability for basic research discoveries in a field to be applied rapidly (in combination with other considerations such as identifiable direct military applications) is a possible reason to restrict IBRC.[121]

The speed with which those scientific discoveries can be commercialized varies dramatically between fields. Discoveries in some fields (e.g., pure mathematics) often have little alignment with possible product applications and can take a highly variable amount of time to come to fruition. Conversely, in fields like cancer research and semiconductor physics, research and applications are more closely connected, although both include curiosity-driven basic research conducted without an application in mind. Progress in fields ranging from the life sciences to AI is accelerating some kinds of translational work and making it harder to separate basic from more applied research in some domains.[122] Advances in communica-

---

118 For a prominent case study, see James D. Watson, *The Double Helix: A Personal Account of the Discovery of the Structure of DNA*, Touchstone, 2001.

119 Ashton B. Carter, "Fundamental Research: Memorandum for the Secretaries of the Military Departments," U.S. Department of Defense, May 24, 2010.

120 Chandra Nath Saha and Sanjib Bhattacharya, "Intellectual Property Rights: An Overview and Implications in Pharmaceutical Industry," *Journal of Advanced Pharmaceutical Technology & Research*, Vol. 2, No. 2, 2011.

121 National Academy of Engineering, *Scientific Communication and National Security*, National Academies Press, 1982, p. 5.

122 Working Group on Science and Technology in U.S.-China Relations, *Meeting the China Challenge: A New American Strategy for Technology Competition*, UC San Diego School of Global Policy and Strategy, November 16, 2020.

tions technology, including online publication and dissemination of results and the ability of remote collaborators to access data, is accelerating knowledge creation and diffusion in many fields.

4.2.9. Amount of Collaboration-Related Overhead

IBRC participants must navigate the regulations and bureaucracies of each partner in addition to establishing the terms of the collaboration itself.[123] For example, an NSF interviewee indicated that the European General Data Protection Regulation, which affects communication of personal data across borders, has caused challenges for U.S. funders and researchers. Language and cultural differences are other forms of overhead.

Both the actual and anticipated overhead can influence collaboration decisions because researchers are unlikely to attempt to initiate collaborations that they believe will be more trouble than they are worth. Uncertainty can be particularly damaging. Multiple interviewees indicated that they or others had decided not to pursue international collaborations due to concerns that the projects could result in bad press for their organization, protracted legal quagmires, or a reduced chance of their projects being funded.

By changing the overhead associated with IBRC with particular partners, government funders have the ability to influence collaboration decisions and attract top researchers. For example, NSF interviewees indicated that one reason NSF staff work with foreign counterparts to create joint funding calls is to reduce the burden associated with finding funding by allowing a whole team to apply for funding using a single application and receive a single decision.

While ANR is optimizing its processes for collaborations with EU programs, both NSF and DoD interviewees remarked that collaborating with EU entities can be complex, which discourages engagements.

Chapter Five discusses how collaboration overhead sometimes challenges IBRC at DoD.

4.3. Metrics and Data Sources to Identify and Evaluate International Basic Research Collaboration Opportunities

This section presents a variety of example metrics suggested by the literature and interviewees that might be used to characterize the considerations discussed in the previous section, including many that DoD program managers (including DoD IOPs) currently use when making funding decisions about proposed IBRC projects.

Table 4.2 lists example metrics relevant for considering potential partners. The metrics examine what a partner might contribute, whether the potential partner has the intent and

[123] The project did not examine which types of overhead are particularly burdensome or identify strategies that researchers and their institutions use to manage the associated challenges.

TABLE 4.2
Example Metrics for Considering International Basic Research Collaboration with a Specific Partner

Consideration	Example Metric(s)	Notes
Access to unique and diverse resources	• Quantity of publications • Quality of publications • Access to unique facilities, materials, equipment, etc.	• Considers ability to contribute expertise, facilities, materials, equipment, etc.
Ability to combine resources and share the burden	• Ability of partner to leverage past foreign investments that created data, expertise, equipment, etc. • Ability of partner to contribute resources	
Build and strengthen relationships	• SME assessments of long-term value of relationship • SME assessment (based on bibliometrics, network analysis, domain knowledge) of importance of location of partner in scientific network to DoD • SME assessment of ability to further strategic leadership goals	• Relationships that will provide long-term benefits and augment DoD's current network are of high value
Trust in a partner	• Partner reputation for protecting information and complying with research norms • SME assessment of partner capabilities and alignment of capabilities with other partners' interests	• Data could come from SMEs and the intelligence community • This consideration looks at both the intent and capability of partners to misuse research results
Amount of collaboration-related overhead	• SME assessments of time needed to complete each partner's processes	• Overhead can depend on the field and the partner • Compiling data on overhead associated with working with different partners and topics could guide efforts to accelerate processes

NOTE: SME = subject-matter expert.

capability to act in a way that is not in the best interests of others on the team, and the amount of overhead their inclusion would add. Many of the metrics, such as those on quality and quantity of publications could be tailored to consider an individual researcher, an organization (e.g., a university), a country, or a group of countries (e.g., NATO). When considering an individual researcher, the benefits the person might bring to a collaboration naturally focus on the scientific field in which the person works. When considering organizations, countries, and groups of countries, there are depth versus breadth trade-offs between considering scientific fields separately and looking at the overall scientific strengths of an entity.

Table 4.3 lists example metrics relevant for considering ways that IBRC might be able to advance progress in a scientific field in view of the risks associated with that field. For many

TABLE 4.3
Example Metrics for Considering International Basic Research Collaborations in a Scientific Field

Consideration	Example Metric(s)	Notes
Ability to combine resources and share the burden	• Costs of infrastructure, data generation, etc. • SME assessments of potential for collaboration to address shared mission (e.g., future interoperability) with partner	• High costs suggest burden-sharing potential, either through co-funding infrastructure (rare for DoD) or seeking partners with infrastructure access (more common for DoD)
Economic impacts	• Prevalence of IP theft in field • SME assessment of potential of collaboration to strengthen defense industrial base and supply chains in desirable (undesirable) locations	• Data has implications for which partners are preferable
Defense impacts	• SME assessments of importance and sensitivity of field to DoD • Presence of export controls • Presence of field on leadership priority lists	• Data has implications for which partners are preferable
Maturity of a scientific field	• SME assessment of state of field	
Speed of transition within a scientific field	• SME assessments of existence of well-defined commercialization or transition paths	

of the field-specific considerations, the results have implications about which partners are preferable. For example, in fields with strong defense or economic impacts, partners with reputations for trustworthiness with whom relationships will advance leadership goals are preferred. For some metrics, such as the cost of infrastructure in a field, large values are associated with high collaboration potential. For other metrics, such as the prevalence of IP theft in a field, low values are associated with higher collaboration potential.

The example metrics identified draw on a range of data sources, some more readily available than others. As Chapter Five discusses in more detail, one challenge for DoD IBRC is fragmentation of the associated data sources, particularly those that could be used to identify and evaluate collaboration opportunities.

Multiple types of SMEs have relevant expertise. Researchers and program managers (including DoD IOPs) are often SMEs on the resources that are most important for making progress in their fields and who has access to those resources. DoD IOPs gain additional expertise from visiting laboratories of foreign researchers and attending workshops in the region where they are stationed. In turn, DoD IOPs often draw on the expertise of DoD intramural laboratory personnel and other service program managers to assess the value of proposed work. Both researchers and the intelligence community might have insights into the trustworthiness of potential partners. For example, someone in a researcher's professional network might have worked with a potential partner in the past and have firsthand knowledge of the person's trustworthiness. Similarly, the intelligence community might have knowledge of activities of the potential partner's institution and country. Security, legal, and policy experts often understand the potential defense and economic impacts of proposed collaborations, what laws and regulations allow, and what leadership would support.

Each of the Reliance 21 Communities of Interest, cross-service groups that coordinate strategies within specific S&T areas across DoD, are tasked to understand DoD interests within their topics of responsibility (across the full spectrum of TRLs).[124] Under the 2014 DoD *International S&T Engagement Strategy*, the Reliance 21 Communities of Interest were also asked to understand the associated international S&T landscape and use that information in identifying and prioritizing engagement gaps and opportunities.[125] The 2020 *Department of Defense International Science and Technology Engagement Strategy* acknowledges the expertise of the Reliance 21 Communities of Interest but suggests that a forum dedicated to sharing information and planning would be more useful.[126]

USG organizations have sponsored a range of relevant analytic efforts.[127] For example, the ARO has sponsored data analytics efforts to identify research areas of importance to the Army where researchers outside of the United States are working at the leading edge.[128] AFOSR has also sponsored data analytics work to identify emerging basic research topics.[129] NSF interviewees indicated that NSF relies on a combination of data analytics and knowledgeable program managers to understand which collaborations would be most worthwhile to facilitate.

In order to comply with Section 1286 of the FY 2019 National Defense Authorization Act ("Initiative to Support Protection of National Security Academic Researchers from Undue Influence and Other Security Threats"), DoD started requiring additional information about all key personnel on funding opportunities, including key personnel that DoD does not fund.[130] For example, applicants must provide information about all current and future support, including the funders, amount of funding, and level of commitment.

---

[124] DoD, *Department of Defense International Science and Technology Engagement Strategy: A Unified Approach to Strengthen Alliances and Attract New Partners*, October 2020.

[125] DoD, *International S&T Engagement Strategy*, July 2014.

[126] DoD, *Department of Defense International Science and Technology Engagement Strategy: A Unified Approach to Strengthen Alliances and Attract New Partners*, October 2020.

[127] The project did not attempt to catalog all DoD and USG data sources and analytic efforts relevant to IBRC.

[128] ARO, *2019 ARO in Review*, U.S. Army Combat Capabilities Development Command, Army Research Laboratory, undated, p. 18.

[129] AFRL, AFOSR, *AFOSR International Office FY19 Annual Report*, undated, p. 20.

[130] Michael D. Griffin, *Actions for the Protection of Intellectual Property, Controlled Information, Key Personnel and Critical Technologies*, Memorandum for Under Secretary of Defense for Acquisition and Sustain-

The amount of collaboration-associated overhead is one of the considerations discussed in the section above. Readily available analysis identifying collaboration opportunities that would allow U.S. researchers to contribute to cutting-edge discoveries that would not otherwise be possible would likely be well received. Time-consuming processes, particularly those that necessitate manually gathering data from multiple sources to justify the value of a proposed collaboration, would likely discourage IBRC.

ment, Assistant Secretary of Defense for Acquisition, Service Acquisition Executives, Special Operations Command, March 20, 2019.

CHAPTER FIVE

Present and Future Challenges to International Basic Research Collaboration Success at the U.S. Department of Defense

This chapter first looks at how DoD and other organizations measure IBRC success and reiterates that a key reason that DoD supports IBRC is to strengthen its network of relationships. The chapter then discusses three factors that reduce the ability of DoD to identify and establish basic research relationships: (1) fragmentation of data, (2) the sentiment among some researchers that DoD is not a partner of choice, and (3) the reluctance of some DoD researchers to engage with foreign researchers. The chapter ends by discussing some trends and their implications for the future strength of DoD basic research.

5.1. How the U.S. Department of Defense and Others Measure International Basic Research Collaboration Success

Almost all of the interviewees, regardless of affiliation, commented on the well-known challenges of measuring the success of basic research investments and indicated that measuring IBRC success presents similar difficulties. Table 5.1 shows a sample of the metrics the project's four example organizations use to measure IBRC success. Additionally, NSF's National Center for Science and Engineering Statistics publishes multiple reports, including the Science and Engineering Indicators series, that look more broadly at a range of topics including U.S. and international trends.[1]

Both ANR and NSFC, but none of the U.S. organizations studied, publish the fraction of their budget that supports projects with international partners.[2,3] This is likely because many U.S. collaborations are initiated in a bottom-up manner and take place outside of a call dedicated to international collaboration. An NSF interviewee indicated that NSF is increasing its

---

1 NSF, "Science and Engineering Indicators," webpage, undated.

2 ANR, "Transnational Collaborations," webpage, undated.

3 NSFC, *2018 Annual Report*, English version, undated, pp. 5–6, 126.

TABLE 5.1
Sample Metrics for Measuring International Basic Research Collaboration Success

Example Organization	Sample Metrics
ANR	• Number of awards of various types that include international partners • Amount of funding that supports French participation in international joint research projects • Number of EU calls that it supports
DOE	• Number of publications, discoveries, advancements, and presentations from international collaborations
NSF	• Volume of publications from international collaborations • Quality of publications from international collaborations • High-profile awards from international collaborations
NSFC	• Number of awards of various types that include international partners • Number of international agreements established • Amount of funding that supports Chinese participation in international joint research projects

SOURCES: NSF, "About the National Science Foundation," webpage, undated; NSB, NSF, *Publications Output: U.S. Trends and International Comparisons*, NSB-2020-6, December 17, 2019; NSFC, *2018 Annual Report*, English version, undated; NSFC, *International Cooperation Bureau (List of NSFC-Signed Cooperation Agreements [MOUs])* [NSFC签署合作协议(谅解备忘录)情况一览表], March 3, 2020; ANR, "Transnational Collaborations," webpage, undated; a DOE interviewee.

analytic capability in order to understand the scale and nature of international collaboration among its awardees. That capability would enable NSF to better analyze the information on planned international collaborations submitted as part of proposals in order to understand the nature of the collaborations (e.g., attending a foreign conference, partnering on the project) and the topics and locations on which collaborations are occurring and why.

DoD IOP interviewees identified two main ways the services attempt to determine the success of their grants to foreign PIs. The first method is by quantifying the traditional research outputs, such as numbers of presentations, publications, publications in prominent journals, and patents. The second approach is to track transitions. As discussed previously, one of the functions of IOPs is to create conduits between basic and applied research. The resulting transitions and knowledge transfers can take multiple forms. For example, the work could advance in TRL, or it could transition to a service laboratory with the service laboratory funding subsequent efforts. Other metrics include press releases, which can be used to illustrate the extent to which DoD is the partner of choice. In order to maintain the connections, DoD follows its foreign grant awardees throughout their careers and notes significant accomplishments such as major grants or international awards.

The *AFOSR International Office FY19 Annual Report* identified several more metrics:

- the amount of funds that the Air Force's applied research organizations co-invest with the International Office, which indicates the degree to which the Air Force's interna-

tional basic research investments are integrated with the Air Force's applied research portfolio
- the creation of long-term relationships with former awardees who move into positions of influence
- opportunities to engage in science diplomacy resulting from IOP awards or in-region presence.[4]

The project team did not take an in-depth look at factors that correlate with collaboration success. Both NSF and DOE interviewees, however, did comment on the importance of the support of the U.S. scientific community to collaboration success, and both organizations attempt to ensure that ground-level support exists before starting new large-scale, international collaborations or issuing joint calls for proposals with counterparts. DoD IOP interviewees indicated that pre-grant interest and demand from DoD's intramural labs increases the chance of a project successfully transitioning. Additionally, the project summarized literature on factors that contribute to success in international research collaborations (Table 5.2 and Appendix D).

5.2. Current Challenges to International Basic Research Collaboration Success at the U.S. Department of Defense

This section discusses three challenges to IBRC success at DoD. The first is the lack of efficient and effective ways for IOPs who are working primarily top-down to assemble the information needed to identify and evaluate collaboration opportunities. The second challenge

TABLE 5.2
Success Factors and Inhibitors of International Research Collaborations

Success Factor	Associated Inhibitor
Establishing terms of engagement early	Not establishing terms of engagement
Free, open, and clear communication with cultural understanding	Communication barriers (cultural, bureaucratic delays)
Effective leadership	Ineffective or no leadership
Diversity of participants	Lack of diversity
Sufficient and flexible funding	Insufficient or inflexible funding
Knowledge of relevant regulations and processes	Opaque bureaucratic environment
Regulations facilitating researcher mobility and information exchange	Regulatory environment that unduly prohibits mobility and information exchange

SOURCE: RAND compilation of information discussed in Appendix D.

4 AFRL, AFOSR, *AFOSR International Office FY19 Annual Report*, undated, p. 5.

is the preference of some researchers not to work for DoD, and the third is the preference of some DoD intramural researchers to avoid engaging internationally. The second and third challenges reduce DoD's ability to (1) form the relationships that will provide a solid foundation for the organization's research efforts and (2) transition basic research discoveries to the defense research enterprise.

5.2.1. Fragmented Resources for Identifying and Evaluating Collaboration Opportunities

As discussed in Section 4.3, a variety of people have knowledge relevant to identifying and evaluating DoD IBRC opportunities, and multiple organizations are conducting relevant analysis. Coordination activities with different levels of formality take place on a variety of scales. At the formal, USG-wide level, the OSTP advises the President on S&T issues and leads efforts to develop and coordinate USG research strategy, budgeting, and policy.[5] OSTP includes the National Science and Technology Council (NSTC), which coordinates executive branch S&T investment.[6] Both OSTP and NSTC's structures and attention to international concerns vary depending on the administration. NSTC processes involve convening and coordination, not control, and agency willingness to participate varies. Additionally, project interviewees identified multiple forms of ad hoc coordination and information-sharing among USG personnel (e.g., DoD IOPs coordinating with Department of State–led country teams in U.S. embassies, DoD IOPs advising NSF personnel on international researchers with whom NSF personnel might meet during MULTIPLIER visits, DoD personnel sharing information with each other).

Interviewees identified multiple challenges with current coordination and data management efforts:[7]

- Information is shared manually, and sharing relies largely on personal connections. Thus, matchmaking by IOPs between international and DoD researchers and determining with whom different pieces of information should be shared is difficult.[8]
- IOPs typically hold their positions for only a few years before rotating into new jobs. With each rotation, knowledge and expertise are lost.
- IOPs cannot effectively review the world's growing basic research portfolio and bring high-value opportunities to the attention of the right DoD personnel. In order for the current number of IOPs to collectively cover all research topics worldwide, each has a

5 Public Law 94-282, National Science and Technology Policy, Organization, and Priorities Act of 1976.

6 William J. Clinton, "Establishment of the National Science and Technology Council," Executive Order 12881, November 23, 1993.

7 One to two interviewees provided each example.

8 The project did not analyze the extent to which the lack of integrated data sources impacts intramural and extramural DoD researchers.

broad portfolio. IOPs are most successful at identifying the best in the world in the areas of their expertise. In other areas, they can provide too many leads for others to screen.
- Documentation, such as trip reports from IOPs and researchers, is not stored in a searchable, central repository.
- Existing DoD databases of grants and international agreements are not integrated with other resources.
- The completeness of the data that DoD and USG research organizations compile on the amount of U.S. funding that supports U.S. participation in international collaborations varies by funding archetype. (Figure 3.2 describes funding archetypes.) In particular, data for the uncoordinated funding for joint projects archetype is sparse.[9]
- DoD and USG research organizations do not appear to engage in large-scale, systematic data collection and mining.

In contrast, China has embraced systematic, big-data approaches. The U.S. Senate Permanent Subcommittee on Investigations found that China systematically targets critical technologies and directs its international engagement accordingly.[10] A study by the Australian Strategic Policy Institute identified more than 600 Chinese overseas talent recruitment stations that work to identify and recruit foreign experts.[11] The same study found that Chinese groups had also assembled large databases on foreign scientists to support their recruitment efforts.[12] As discussed elsewhere in this report, the deception inherent in many of China's foreign engagement practices is contrary to standard U.S. research norms. Nonetheless, there could be opportunities for the USG to overtly consolidate its data and analysis into resources that could allow departments to better leverage each other's efforts. For example, one DoD IOP interviewee suggested that knowledge management tools that provide visibility into contacts and engagements and make resources such as IOP site visit reports and DoD intramural researcher trip reports easily discoverable would increase productivity. The interviewee elaborated that visibility from being able to use such tools across multiple services and ideally across multiple USG departments would assist with coordination.

---

9 An interviewee indicated that NSF is increasing its analytic capability to more fully analyze the information on planned international collaborations that is submitted with proposals that are responding to calls not dedicated to international collaborations.

10 U.S. Senate Permanent Subcommittee on Investigations, "Threats to the U.S. Research Enterprise: China's Talent Recruitment Plans," staff report, 2019.

11 Alex Joske, *Hunting the Phoenix: The Chinese Communist Party's Global Search for Technology and Talent*, Australian Strategic Policy Institute: International Cyber Policy Centre, No. 35, August 2020.

12 Alex Joske, *Hunting the Phoenix: The Chinese Communist Party's Global Search for Technology and Talent*, Australian Strategic Policy Institute: International Cyber Policy Centre, No. 35, August 2020.

5.2.2. Preference of Some Researchers to Work for Research Funders Other Than the U.S. Department of Defense

Top researchers have multiple collaboration opportunities available to them and consider a range of factors when deciding where to focus their efforts. Factors include the overhead, flexibility, and prestige associated with research funds and the potential future opportunities that might result from the work. The relative weight of the factors varies among researchers.

Multiple DoD interviewees discussed the importance of DoD (and the United States) being the partner of choice.[13] By engaging internationally, the United States can reduce the chances of other countries, such as strategic competitors, becoming the partner of choice. DoD IOP interviewees considered such efforts to be particularly critical in parts of the world, such as South America, where the United States is a comparatively recent partner and collaboration patterns have not yet been set. Interviewees feared that if DoD does not engage, it would leave an opening for another country to become a preferred partner and reap the associated benefits. As discussed in Section 3.2.3, although China is often more generous, many international academics find the prestige and favorable IP terms of DoD grants attractive.

Countering DoD's attractive characteristics, DoD's connection to defense deters some researchers (including some U.S. researchers). Multiple interviewees indicated that attitudes and policies about accepting defense funding or conducting dual-use research vary among countries, institutions, and individuals.[14] For example, France is generally open to looking at defense and civilian research holistically.[15] Similarly, researchers educated in the United States as well as researchers in countries where research funding is particularly scarce are often more open to accepting DoD funding. In contrast, as part of the legacy of World War II, Japanese and German researchers are often particularly reluctant to do anything that might be perceived as mixing defense and civilian research.

DoD IOPs interviewees identified multiple approaches for overcoming researchers' initial reluctance to engage with DoD:[16]

- In cases where the prohibition is on accepting funding for defense applications rather than accepting defense funding, IOPs can explain that the funding would be for basic research and that basic research has a range of applications, including civilian applications.
- Including trusted partners can reassure academic PIs who might otherwise decline to accept DoD funds. For example, involving a third country with strong, preexisting ties to both DoD and the other country can be helpful.

---

[13] One to three interviewees provided each illustration.

[14] One to two interviewees provided each example.

[15] Christian Mölling, *Defense Innovation and the Future of Transatlantic Strategic Superiority: A German Perspective*, German Marshall Fund of the United States, March 23, 2018.

[16] Unless attributed to another source, one to two interviewees provided each example.

- Partnering with defense organizations in a foreign researcher's country (e.g., sharing information about interests and prospective and current grantees, coordinating awards), can help to reassure foreign researchers that their government is okay with a proposed activity. Such partnering also has the added benefit of giving DoD insight into the foreign organization's research interests.

Multiple DoD IOP interviewees also indicated that the amount of administrative work associated with DoD grants can discourage some researchers from working with DoD.[17] For example, some prospective awardees opt not to pursue DoD grants due to the effort that would be required to work through their institutions' bureaucracies to gain approval to accept defense funding. Others consider the standard U.S. "grants.gov process" too burdensome to be worthwhile for small grants. (One DoD IOP, however, indicated that limiting the grant size is one way to mitigate risk.) Lack of administrative personnel at some foreign universities who speak English well enough to assist with the paperwork can also be problematic. Delays in reimbursing Collaborative Science program participants for travel expenses can unduly stress the personal finances of some foreign researchers. Delays in obtaining approvals for researchers to work on the Army Research Laboratory open campus have resulted in researchers to moving on to other opportunities.

5.2.3. Reluctance of Some U.S. Department of Defense Researchers to Engage with International Colleagues

As one of DoD's IBRC goals is to connect foreign researchers to the intramural DoD network, the willingness and ability of DoD intramural researchers to collaborate internationally is critical to DoD's IBRC success. Interviewees, however, mentioned two main categories of concerns that reduce the willingness of some researchers who work in DoD intramural laboratories, to engage internationally: (1) the time and effort needed to initiate and conduct IBRCs and (2) the potential for unintentionally violating rules or causing a security breach. (The concerns, particularly the first one, are also relevant to academic researchers, regardless of whether they use DoD funds.[18])

As discussed in Chapter Four, the amount of collaboration-related overhead is a consideration when determining whether to pursue an IBRC opportunity.[19] Requirements to report interactions with foreign nationals deter some DoD intramural researchers from engaging with foreign researchers. Recent pushes to require researchers to establish nondisclosure agreements to cover discussions with foreign colleagues also discourage interactions. Security reviews of foreign collaborations, particularly with new partners, can take considerable

17 One to three interviewees provided each illustration.

18 The project did not compare attitudes toward international engagement among DoD intramural researchers, DoD-funded grantees, or U.S. basic researchers supported by other parts of the USG.

19 One to two interviewees provided each example.

effort and can cause uncertainty about if or when a project will be allowed to start. DoD intramural laboratory dislike for purchasing foreign research tools can be problematic in cases in which there is no U.S. substitute, as in the case of computational materials science codes. The collective impact of overhead can be DoD researchers opting to work with domestic "B-teams" rather than international stars.

Concerns (whether justified or not) about unintentionally violating rules and the associated consequences can reduce the willingness of DoD researchers to participate in IBRC, with intramural and extramural researchers more worried about security violations and potential legal quagmires, respectively.[20] Lack of clarity about what types of foreign interactions are unacceptably risky, which types are acceptable as long as they are reported, and what constitutes a foreign interaction that needs to be reported can have a chilling effect on outreach to international colleagues.[21]

5.3. Trends and Their Implications for Future Success

This chapter concludes by briefly discussing two trends and their implications for DoD: (1) the decreasing share of the world's basic research being conducted in the United States and (2) the increasing reliance of the defense sector on ideas generated in the civilian sector. Both increase the amount of basic research activity that DoD IOPs need to monitor and the number of international collaboration opportunities they need to consider. Thus, going forward, it would likely be advantageous for DoD IOPs to be able to identify, evaluate, and act on a larger number of IBRC opportunities rapidly.

5.3.1. The Portion of the World's Basic Research Conducted in the United States Is Decreasing

The research enterprise is becoming increasingly global, with the share of the world's R&D that the United States funds decreasing from 69 percent in 1960 to 28 percent in 2016.[22] The Defense Science Board (DSB) identified growth in extranational basic research as a concern in 2012, and the concern has grown since then.[23] In 2009, NRC cautioned that

[20] One to two interviewees provided each example.

[21] The project did not examine the impact of the matchmaking functions of IOPs on overcoming the reluctance of some DoD intramural researchers to engage internationally or whether other approaches would be useful.

[22] John F. Sargent, Jr., *Global Research and Development Expenditures: Fact Sheet*, Congressional Research Service, June 27, 2018. The percentages refer to all R&D, not just basic research.

[23] DSB, *Report of the Defense Science Board Task Force on Basic Research*, Office of the Under Secretary for Defense for Acquisition, Technology, and Logistics, January 2012.

> American security and prosperity now depend on maintaining active engagement with worldwide developments in science and technology, and with the global economy. While the United States remains a world leader in advanced science and technology, it no longer dominates; it is now *among* the leaders. We are increasingly interdependent with the rest of the world.[24]

As one former DoD research official observed in 2020, when it comes to basic (and other) research, the world is multipolar, and "no one nation is likely to assume a globally dominant majority role in the foreseeable future."[25]

The increase in basic research and quality basic research outside of the United States increases the amount of basic research activity that the DoD research enterprise needs to track. Thus efficient processes that can be scaled to allow DoD IOPs to keep up with the growing amount of quality international research will likely become increasingly valuable going forward. Additionally, analyzing the effectiveness and extent of the efforts that DoD intramural and extramural researchers use to identify IBRC opportunities, which are presumably largely bottom-up, as well as increasing visibility into the IBRCs they are pursuing, could provide a fuller understanding of DoD's overall IBRC portfolio and approaches.[26]

5.3.2. The Flow of Ideas from the Civilian to the Defense Sector Is Increasing

Whereas the commercial sphere once benefited from defense innovations in a trickle-down manner, the flow has been reversed for several decades. NRC observed that

> By 2000, except for small but important niches in the military sector, components for both military applications and the commercial market were drawn from the same technology base. The key differentiating characteristic in the military market now was not the underlying scientific or technological information, components, or products, but the methods by which these were applied to specific military uses.[27]

---

[24] NRC, *Beyond "Fortress America": National Security Controls on Science and Technology in a Globalized World*, National Academies Press, 2009, p. 2.

[25] Melissa Flagg and Paul Harris, "System Re-Engineering: A New Policy Framework for the American R&D System in a Changed World," Center for Security and Emerging Technology, September 2020.

[26] The approaches that DoD researchers use to identify and evaluate opportunities (e.g., reaching out to their networks) might be more robust to increases in basic research volume than some that DoD IOPs currently use.

[27] NRC, *Beyond "Fortress America": National Security Controls on Science and Technology in a Globalized World*, National Academies Press, 2009, p. 21.

In 2012, DSB amplified that

> To aggravate the situation, most of the scientific work now done in the United States lies outside DoD's purview and, thus, DoD no longer has access to much of the nation's best and brightest science talent as it did during the Cold War.[28]

Beginning in the 2010s, DoD launched new efforts intended to engage nontraditional sources of technology for defense. For example, through the Defense Innovation Board, the Defense Innovation Unit, and the National Security Innovation Network, DoD worked to build connections between defense entities and start-ups and academia. The *Summary of the 2018 National Defense Strategy of the United States of America* reaffirmed that the commercial sector will generate many future defense technologies.[29]

As basic research outside the United States increases, the amount of basic research that DoD IOPs need to monitor also increases. An additional implication is that the reluctance of some researchers to engage with DoD could become more problematic.

---

[28] DSB, *Report of the Defense Science Board Task Force on Basic Research*, Office of the Under Secretary for Defense for Acquisition, Technology, and Logistics, January 2012, p. xi.

[29] Jim Mattis, *Summary of the 2018 National Defense Strategy of the United States of America: Sharpening the American Military's Competitive Edge*, U.S. Department of Defense, undated.

CHAPTER SIX

Conclusions

This chapter gathers keys observations from the study and offers some suggestions for DoD to improve how it uses IBRC.

6.1. Observations

The project team identified five main observations:

Observation 1: Including international basic research collaborators within its network is one way DoD seeks to support its mission. Benefits of IBRC include reducing technological surprise, participating in cutting edge discoveries, leveraging investments of partners and allies, accessing diverse resources, and integrating international scientific thought leaders into DoD networks. While strategic considerations add to or detract from the scientific benefits for some collaborations, declining to participate in IBRC is not risk-free and carries costs in the form of lost opportunities.

Observation 2: Adversaries who seek to exploit traditional U.S. research norms challenge the efforts of U.S. organizations to form a solid basic research foundation. As discussed in Section 1.2 (Role of Research Integrity in International Basic Research Collaboration), China and others have violated those norms to gain an advantage. The United States, including DoD, would benefit from being able to compete fairly and effectively with other research funders for the best researchers in the world. The OSTP, U.S. research funders including DoD, university associations, and others have worked to combat the problem by increasing awareness and requiring additional disclosures of potential conflicts of interest and commitment. As DoD has observed, additional policies to clarify how to use the information disclosed to assess risks and make decisions are needed.[1] The absence of clear direction could allow undesirable conflicts to persist provided they are declared and, as one NSF interviewee suggested, could also needlessly dissuade U.S. researchers from proposing to work with top international collaborators out of fear that the association could reduce their chances of being awarded funds.

---

[1] U.S. Government Accountability Office, *Federal Research: Agencies Need to Enhance Policies to Address Foreign Influence*, GAO-21-130, December 2020, p. 39.

Observation 3: DoD is unable to capitalize on some IBRC opportunities due to some DoD intramural researchers' reluctance to engage in IBRC and some U.S. and foreign researchers' preference not to accept U.S. or defense funds. Researchers decline DoD funding for a variety of personal reasons and institutional and country-specific policies. Interviewees indicated that reluctance among DoD intramural researchers has multiple sources, including (1) concerns about unintentionally violating rules (e.g., U.S. ITAR, which, although they apply only rarely to basic research, can cause challenges if a project unexpectedly becomes applied) or causing a security breach and (2) the time and effort needed to initiate and conduct IBRCs. Collaboration-related overhead, including the time needed to navigate cultural and language barriers and obtain approvals and funding through multiple bureaucracies, can discourage both DoD and non-DoD researchers from IBRC and offset the benefits of collaborating, particularly in fast-moving fields. In view of the regional emphases of some countries (e.g., see Section 4.2.3.2, which describes how ANR prioritizes French-German collaboration in support of the Franco-German Treaty of Aachen) and the specialized funding processes and regulations associated with some regions (e.g., the EU), it might become increasingly difficult for the USG to support researchers' efforts to maintain scientific ties worldwide.

Observation 4: DoD's approach of stationing IOPs overseas to identify, engage with, and directly fund foreign basic researchers is distinctive. Few other countries have the resources to invest in proactively seeking IBRC opportunities or influencing research networks and directions worldwide. China also conducts extensive international outreach but attaches far more substantial constraints and conditions to the funds it provides, such as transferring IP from U.S.-funded work to China and hiding the affiliation with China.[2] Most other organizations and countries do not regularly fund foreign researchers working in the researchers' home countries.

Observation 5: As basic research continues to internationalize and an increasing proportion of ideas originate outside of defense research institutions, DoD would benefit from being able to rapidly identify, evaluate, and act on a growing set of IBRC opportunities. Current tools, approaches, and staffing levels of the DoD IOPs might be insufficient to keep up with the demand.

6.2. Suggestions to Help the U.S. Department of Defense Improve How It Uses International Basic Research Collaboration

The project team identified five suggestions that might help DoD improve how it uses IBRC:

Suggestion 1: Provide knowledge management tools to help DoD research funders and DoD intramural researchers identify, evaluate, establish, and conduct international col-

2 U.S. Senate Permanent Subcommittee on Investigations, "Threats to the U.S. Research Enterprise: China's Talent Recruitment Plans," staff report, 2019.

laborations. For example, an IOP interviewee indicated that such tools would ideally make a variety of resources easily discoverable and usable such as IOP site visit reports, international trip reports from DoD intramural researchers, results from service studies that analyze the international landscape, and collaboration guidance. Tools could also provide visibility into contacts and engagements. (See Section 5.2.1.) Other useful features might include partner-centric and scientific-field–centric views—perhaps including data for some of the metrics presented in Tables 4.2 and 4.3—that could help personnel assess a range of scientific and strategic considerations. DoD intramural researchers could use such tools to augment their own existing knowledge about their fields and scientific networks to scope and prioritize bottom-up opportunities. Such tools could also help DoD IOPs working top-down to identify and evaluate leads for engagement. Because basic research is relevant to a range of applications, a DoD IOP interviewee suggested that a single platform that allows users to leverage data from across DoD (and ideally the USG) would be preferable to service-specific implementations. Including information from other USG departments (while continuing to respect preferences of researchers who do not wish to do defense work) could increase the utility of the tools and help DoD to more effectively leverage the investments of other USG departments in understanding the international research landscape.

Knowledge management tools might also collect and provide information to support service implementations of the international S&T portfolio optimization described in the 2020 *Department of Defense International Science and Technology Engagement Strategy*. Conversely, the outputs from that portfolio analysis might be fed into a knowledge management tool and shared more broadly.

Suggestion 2: Review, improve, and publicize policies and guidance to reduce instances of DoD intramural and DoD-funded U.S. academic researchers unnecessarily choosing not to work with high-quality foreign partners. Interviewees, especially those from NSF and academia, expressed concerns that some academic researchers might be avoiding proposing work with some international researchers out of unjustified fears that doing so would reduce their chances of receiving an award. As described in Section 5.2.3, worries (whether justified or not) about unintentionally violating rules discourage some researchers from engaging with international colleagues. Interviewees had several suggestions:

- Better define the level of interaction with a foreign national that security clearance holders must report and when nondisclosure agreements are and are not needed.
- Clarify what types of foreign connections are unacceptable for DoD intramural researchers, research contractors, and basic research grantees; which necessitate additional, case-by-case scrutiny; and how that scrutiny will be conducted.
- Communicate the process that DoD uses to weigh strategic and scientific considerations and to decide whether to engage in fundamental research with a foreign partner on a specific topic.

- Provide additional guidance on how DoD researchers, including academics, should navigate the transition from open fundamental research to more restricted later stage research.

As of October 2021, DoD was updating its policies and processes to address many of the items mentioned in these suggestions. Many DoD components have started using a risk-based framework to consider the sources of support for key project staff, with security personnel often reviewing disclosures (and looking for undisclosed connections) in parallel with the merit review process.[3] Aside from items specifically prohibited by federal law, nothing by itself is necessarily disqualifying.[4] DoD was also participating in an interagency process to finalize guidance for implementing NSPM-33, including requirements for disclosure and research security training.[5] DoD plans to issue policies based on that guidance soon after its release.[6] While DoD has policies on foreign travel and the permissibility of grant or contract funding flowing to foreign entities, these policies are typically specific to individual research offices.[7] Particularly for U.S. academics whose research is supported by multiple services, DoD-wide guidance would be especially useful.

Suggestion 3: Streamline application and approval processes to reduce overhead associated with IBRC. As discussed in Chapter Four, the amount of collaboration-related overhead is one of the considerations that researchers weigh when deciding whether to pursue an IBRC. As discussed in Chapter Five, interviewees indicated that the amount of effort needed for bureaucratic processes can both deter international researchers from applying for DoD grants and can deter DoD intramural researchers from seeking international partners. Interviewees had three associated suggestions for better matching processes to the risk and scale of the proposed activities. First, the USG might create an alternative process to rapidly establish international agreements (or create a process that does not require an international agreement) for government-to-government research collaborations that are comparatively low risk (e.g., relatively small financially, do not produce a product). DoD would likely need new legal authorities for such a process. (See Appendix B for information on the current DoD international agreement process.) While most DoD IBRCs are not government-to-government, such a process might help with projects that both conduct basic research and transition the results. Second, explore options to make the USG-wide "grants.gov process" more flexible and user-friendly for small DoD grants to foreign researchers. Some requirements, such as obtaining a DUNS number, can be particularly challenging for researchers whose administrative personnel might not speak English. Third, consider whether DoD approval processes

3 BRO staff person, email to A. Hottes, October 29, 2021.

4 BRO staff person, email to A. Hottes, October 29, 2021.

5 BRO staff person, email to A. Hottes, October 29, 2021.

6 BRO staff person, email to A. Hottes, October 29, 2021.

7 BRO staff person, email to A. Hottes, October 29, 2021.

should be simplified for IBRC proposals that based on the topic and partner are likely to be of sufficiently low risk as to not merit detailed security analysis. While such an approach might encourage more DoD researchers to consider participating in IBRC, some researchers might still be disinclined to pursue high-benefit, higher-risk collaborations that, although meriting detailed review, are still in the best interests of DoD.

Suggestion 4: Analyze whether adjusting the size, number, and terms of DoD grants to foreign researchers could provide a competitive advantage and help DoD meet its goal of being the partner of choice. As discussed in Section 5.2.2, the comparatively small size of grants that DoD offers to international academics, especially in view of the involved application process (see Suggestion 3), deters some from working for DoD. Interviewees suggested that an analysis of grant practices might also consider (1) whether grant sizes should vary depending on the country of the recipient and (2) the overall amount of basic research funds that DoD should invest in supporting international researchers in view of competing priorities.

Suggestion 5: Assemble a fuller picture of DoD IBRC efforts and processes; in view of the growing quantity and quality of research being conducted outside of the United States and the increased technological competition with the PRC and others, such information could help DoD assess and update its IBRC processes, policies, and strategies. Much of this study focused on the activities of the DoD IOPs for whom IBRC is a core part of their duties. The roles of the much larger number of DoD intramural and DoD-funded U.S. academic researchers in IBRC are more difficult to characterize, particularly because most spend only a small part of their time thinking about or engaged in IBRC. The National Center for Science and Engineering Statistics tracks basic research obligations to foreign performers as part of its *Survey of Federal Funds for Research and Development*. However, that data set does not report on the scale of the USG's obligations to the U.S. side of IBRCs. USG entities have data on the funds they awarded to U.S. researchers through bilateral or multilateral calls for IBRC projects, although public summaries of that information do not appear to be available. More challenging to obtain is information about funds that support the work of U.S. researchers who develop bottom-up collaborations with foreign researchers who acquire funding independently. Bibliometric analyses of funding acknowledgments and author affiliations in published papers are one source of information about collaboration patterns and could be used to approximate the fraction of a researcher's time that was devoted to IBRC. Additionally, interviewees indicated that NSF is working to answer that question and others by more thoroughly analyzing the information it already collects (e.g., grant applications). While some DoD IOPs indicated that consolidated data sources would be useful to support their efforts to understand the overall international basic research ecosystem, particularly in view of the increasing amount of basic research conducted outside of the United States (see Section 5.3.1), the extent to which such resources would be useful to individual DoD researchers, who are more narrowly focused, is not clear. Similarly, while this study identified collaboration-related overhead as a key consideration for many researchers contemplating an IBRC (see Sections 4.2.9

and 5.2.3), it did not determine which types of overhead are most burdensome or the range of strategies researchers and their institutions use to cope with overhead.

In the future, DoD might gather additional information (ideally in a way that minimizes the associated overhead) about how DoD intramural and DoD-funded U.S. academic basic researchers identify and evaluate international collaboration and engagement opportunities, what IBRCs they are pursuing, and how much time and funds they are investing in IBRC activities. DoD might also take a more in-depth look at the challenges that DoD intramural and extramural researchers are encountering and how they are overcoming those difficulties (e.g., what types of overhead are particularly burdensome and what strategies researchers or their institutions use to manage or reduce that burden). Other topics of interest might include the relative effectiveness of the processes that DoD researchers and IOPs use to identify and evaluate IBRC opportunities, the extent to which their efforts complement each other and collectively allow DoD sufficient visibility into—and access to—the basic research ecosystem, the actual and desired relationship between DoD's international and domestic basic research portfolios, and opportunities for improvement.

Overview of Example Organizations

This appendix provides an overview of the IBRC approaches of the project's four example organizations:

- ANR
- NSFC
- DOE
- NSF.

For each organization, the appendix describes ways the organizations support and facilitate IBRC using the three categories introduced in Table 1.1:

- Sponsor research or exploratory activities with an international component.
- Permit and enable the national research enterprise to engage internationally.
- Work with partner governments and international organizations to facilitate IBRC.

See Chapter Two for information on how the project team selected the organizations and Chapter Three for a comparison among the organizations and the U.S. DoD.

A.1. National Research Agency of France

ANR, which was established in 2005, operates under the authority of the French Ministry of Higher Education, Research and Innovation. ANR competitively awards research funds in order to "promote French project-based research and to stimulate innovation by promoting the emergence of collaborative multidisciplinary projects and encouraging collaboration between the public and private sectors."[1] ANR's missions include funding research at a range of TRLs (including basic research), managing government investment programs in higher education and research, and strengthening scientific cooperation with Europe and beyond.[2] ANR money is intended to augment the baseline funding French research organizations

[1] ANR, "Missions," webpage, undated.

[2] ANR, "Missions," webpage, undated.

receive. In 2019, ANR had an operating budget of approximately €40 million and provided €725 million of research funding.[3]

Each year, ANR publishes a Work Programme that describes its research priorities and planned activities for the upcoming year. The 2021 Work Programme contains four components:[4, 5]

- **ANR's Generic Call for Proposals**, which solicits projects led by young researchers as well as three types of collaborations: (1) collaborations among French partners, (2) international collaborative research projects (projets de recherche collaborative—international [PRCI]), and (3) collaborations with commercial entities
- **challenges and calls to respond rapidly to emergencies** (e.g., COVID-19)
- **the "Building the European Research Area (ERA) and France's International Attractiveness" component**, which includes additional bilateral and multilateral calls for proposals
- **the "Economic Impact of Research and Competitiveness" component**, which includes efforts to partner with and transition results to industry.

A.1.1. How ANR Facilitates and Supports IBRC

ANR uses a range of mechanisms to support IBRC (not all activities discussed are exclusively basic research).

A.1.1.1. Funding Research or Exploratory Activities with an International Component

- The PRCI mechanism under the Generic Call solicits proposals for collaborations between French and foreign partners. The foreign partner must be from a country with which ANR has signed a bilateral agreement.[6]
- Activities within the "Building the European Research Area (ERA) and France's International Attractiveness" portion of the Work Programme all have an explicit international element.[7]

---

3 ANR, *Soutenir la recherche dans sa diversité—Funding Research in All Its Diversity: Rapport d'Activité 2019—Annual Report 2019*, undated, p. 72.

4 Other recent Work Programmes are similar.

5 ANR, *Work Programme 2021*, English version, v.1.1, October 22, 2020.

6 ANR, *2021 Work Programme: Generic Call for Proposals: 2021 AAPG Guide: Submission, Evaluation, Selection and Funding Guidelines*, English version, v.1.0, October 2, 2020.

7 ANR, *Accompagnement Spécifique des Travaux de Recherches et d'Innovation Défense: ASTRID* [*Specific Support for Defense Research and Innovation Work: ASTRID*], 2020, p. 13.

A.1.1.2. Permitting and Enabling the National Research Enterprise to Engage Internationally

- Researchers may include foreign partners in non-PRCI applications under the Generic Call provided that each foreign partner has its own funds.[8]
- Some calls that ANR administers for other French funders allow foreign participants. For example, the ASTRID and ASTRID Maturation programs that support dual-use research allow foreign partners subject to the approval of a steering committee.[9]

A.1.1.3. Working with Partner Governments and International Organizations to Facilitate IBRC

- ANR works with foreign funding organization counterparts to establish bilateral and multilateral agreements for cooperative projects.[10]
- ANR engages with multiple international partnerships including the Belmont Forum and the Joint Programming Initiatives to harmonize research agendas and coordinate research funding.[11]
- ANR connects its research priorities to those of other European and international entities.[12]

A.2. National Natural Science Foundation of China

The missions of NSFC include supporting basic research, developing researchers, and promoting economic and social growth.[13] NSFC's 2019 budget was 31.11 billion yuan ($4.52 billion), which it used to support researchers, facilities, and equipment.[14, 15] NSFC's portfolio is

---

8 ANR, *2021 Work Programme: Generic Call for Proposals: 2021 AAPG Guide: Submission, Evaluation, Selection and Funding Guidelines*, English version, v.1.0, October 2, 2020, pp. 11, 12, and 14.

9 ANR, *Accompagnement Spécifique des Travaux de Recherches et d'Innovation Défense: ASTRID* [*Specific Support for Defense Research and Innovation Work: ASTRID*], 2020, p. 13.

10 ANR, *Work Programme 2021*, English version, v.1.1, October 22, 2020.

11 ANR, *Work Programme 2021*, English version, v.1.1, October 22, 2020.

12 ANR, *Work Programme 2021*, English version, v.1.1, October 22, 2020.

13 NSFC, "NSFC at a Glance," webpage, undated.

14 NSFC, "NSFC at a Glance," webpage, undated.

15 Per Exchange-Rates.org (https://www.exchange-rates.org/Rate/USD/CNY/12-31-2018), one dollar was worth 6.8785 yuan on December 31, 2018.

broad and includes mathematics, physics, chemistry, life sciences, earth sciences, engineering, materials sciences, information sciences, management sciences, and health sciences.[16]

When NSFC was established in 1986, it reported directly to China's State Council.[17] In 2018, the Chinese government moved NSFC into MOST, which had previously focused on later-stage research, likely in an attempt to improve the transition of basic research results and China's ability to recruit foreign scientists.[18, 19] Although administratively under MOST, NSFC has officially retained power to operate independently.[20]

A.2.1. How NSFC Facilitates and Supports IBRC

NSFC uses a range of mechanisms to support IBRC.

A.2.1.1. Funding Research or Exploratory Activities with an International Component

- NSFC funds the Chinese side of projects with foreign partners who provide their own resources.[21] In some cases, the projects take place under bilateral and multilateral MOUs. In others, the Chinese and foreign partners develop a letter of agreement during the application process that specifies the project's scope, financial arrangements, duration, and participants as well as IP provisions.
- NSFC also supports personnel exchanges and conferences that take place under bilateral and multilateral MOUs.[22] The personnel exchange and conference funds are intended to enhance the ongoing activities of current NSFC funding recipients.
- NSFC funds young non-Chinese citizens (those who are less than 40 years old) to conduct research in China.[23]

---

16 NSFC, *National Natural Science Fund Guide to Programs 2019*, English version, undated, p. 2.

17 NSFC, "NSFC at a Glance," webpage, undated.

18 Hepeng Jia, "China Reshuffles Science Governance: Moves May Generate Conflict Between Basic and Applied Research Programs," *C&EN*, Vol. 96, No. 18, April 25, 2018.

19 Hepeng Jia, "China's Science Ministry Gets Power to Attract More Foreign Scientists: Science Super-Ministry Gains New Remit to Develop Policies That Benefit International Researchers," *Nature Index*, March 23, 2018.

20 NSFC, "NSFC at a Glance," webpage, undated.

21 NSFC, *National Natural Science Fund Guide to Programs 2019*, English version, undated, pp. 194–199.

22 NSFC, *National Natural Science Fund Guide to Programs 2019*, English version, undated, pp. 199–200.

23 NSFC, *National Natural Science Fund Guide to Programs 2019*, English version, undated, p. 218.

A.2.1.2. Permitting and Enabling the National Research Enterprise to Engage Internationally

- Similar to many other organizations, NSFC allows Chinese and foreign researchers to form teams in a bottom-up manner in the absence of an MOU between NSFC and the organization funding the foreign researchers and apply for NSFC funds for the Chinese side of the collaboration. However, unlike many other organizations where such proposals would compete against proposals that do not include international collaborations, NSFC appears to have a dedicated program to evaluate them.[24] Thus, while researcher-initiated collaborations that take place without an MOU between the funding organizations typically fall in the "Permitting and enabling the national research enterprise to engage internationally" category, those with NSFC fit best in the "funding research or exploratory activities with an international component" category.

A.2.1.3. Working with Partner Governments and International Organizations to Facilitate IBRC

- NSFC establishes cooperative agreements with counterparts to provide frameworks and mechanisms for research cooperation.
- NSFC leadership participates in bilateral and multilateral meetings with counterparts. For example, NSFC organized a 2018 workshop in France that discussed science funding policies and strategies[25] and participated in the 2019 World Conference on Research Integrity.[26] NSFC also conducts cooperation meetings with partners such as Japan,[27] the United Nations Environment Program,[28] and the Research Council of Norway.[29]

---

[24] NSFC, *National Natural Science Fund Guide to Programs 2019*, English version, undated, pp. 194–195.

[25] NSFC, "The International Workshop on Science Funding Strategies and Policies Successfully Held," webpage, undated.

[26] NSFC, "NSFC Vice President Wang Chengwen Attends the 6th World Conference on Research Integrity," webpage, undated.

[27] NSFC, "NSFC Vice President Wang Chengwen Attends the 16th Meeting of China-Japan Committee on Intergovernmental Science and Technology Cooperation and Visits JSPS," webpage, undated.

[28] United Nations Environment Program, International Ecosystem Management Partnership, "UN Environment-NSFC Bilateral Cooperation Meeting," January 29, 2017.

[29] NSFC, "NSFC Secretary-General Han Yu Attends the 94th Meeting of IIASA Council and Visits Norway," webpage, undated.

A.3. U.S. National Science Foundation

NSF is an independent federal agency that supports fundamental research within the fields of science, engineering, and education.[30] Since its establishment in 1950, NSF's mission has been "[t]o promote the progress of science; to advance the national health, prosperity, and welfare; and to secure the national defense; and for other purposes,"[31] with the more recently articulated goal of positioning the United States as "a Nation that is a global leader in research and innovation."[32] Funding of basic research is a fundamental component of NSF's mission. As a 2019 letter from the Office of the Director states, "Basic research is one of the underpinnings of America's ability to sustain its position as an innovation leader, its economic strength, and its national security."[33]

NSF is a funding agency with a workforce of about 2,100 at its headquarters in Alexandria, Virginia.[34] With an annual budget of about $8.5 billion (FY 2021), NSF funds approximately 25 percent of all federally supported fundamental research conducted at colleges and universities in the United States.[35] In a typical year, NSF funding supports about 200,000 scientists, engineers, educators, and students at universities, laboratories, and field sites throughout the world.[36] It additionally supports numerous partnerships between industry, national laboratories, and others. NSF distributes its funding primarily through competitive awards. Every year, NSF receives approximately 50,000 funding requests and makes 11,000 to 12,000 new awards.[37] To ensure fairness and transparency, NSF evaluates proposals using a merit-based review process with two primary criteria: (1) intellectual merit (a project's potential to advance knowledge) and (2) broader impacts (the project's potential to benefit society).[38, 39]

---

30 NSF, *Building the Future: Investing in Discovery and Innovation: NSF Strategic Plan for Fiscal Years (FY) 2018–2022*, NSF 18-45, February 2018.

31 U.S. Code Title 42, Section 1861-75, National Science Foundation Act of 1950, 1950.

32 NSF, *Building the Future: Investing in Discovery and Innovation: NSF Strategic Plan for Fiscal Years (FY) 2018–2022*, NSF 18-45, February 2018.

33 NSF, Office of the Director, *Personnel Policy on Foreign Government Talent Recruitment Programs*, July 11, 2019.

34 NSF, "Who We Are," webpage, undated.

35 NSF, "About the National Science Foundation," webpage, undated.

36 NSF, "Who We Are," webpage, undated.

37 NSF, *Building the Future: Investing in Discovery and Innovation: NSF Strategic Plan for Fiscal Years (FY) 2018–2022*, NSF 18-45, February 2018.

38 NSF, *Building the Future: Investing in Discovery and Innovation: NSF Strategic Plan for Fiscal Years (FY) 2018–2022*, NSF 18-45, February 2018.

39 NSF, "About the National Science Foundation," webpage, undated.

A.3.1. How NSF Facilitates and Supports International Basic Research Collaboration

NSF uses a range of mechanisms to support IBRC, including both top-down and bottom-up approaches.

A.3.1.1. Funding Research or Exploratory Activities with an International Component

NSF runs multiple activities specifically designed to foster international collaboration. For example,

- NSF and foreign funding agencies issue multiple joint calls inviting teams of U.S. and foreign researchers to submit applications for support. For example, the Dimensions of Biodiversity program is open to teams from the United States, China, Brazil, and South Africa.[40] Similarly, the Partnerships for International Research and Education (PIRE) program, which is currently on hold to allow the solicitation to be updated, supports projects where international collaboration is required for advances in research and education.[41]
- Through the Accelerating Research Through International Network-to-Network Collaborations (AccelNet) program, NSF supports the creation of international research networks that address grand challenges.[42] Rather than directly supporting research, the AccelNet program funds activities such as sharing resources, knowledge, and expertise to enhance network formation and strength.
- The International Research Experiences for Students program supports international research and research-related activities for U.S. science and engineering students by focusing on, "active research participation by undergraduate and/or graduate students in high quality international research, education and professional development experiences in NSF-funded research areas."[43]
- The MULTIPLIER program deploys NSF experts to selected international regions to build networks and explore opportunities for science and engineering collaboration.[44] MULTIPLIER expeditions focus on fields of science and engineering where interna-

---

40 NSF, "Funding: Division of Environmental Biology—Dimensions of Biodiversity," webpage, undated.

41 NSF, OISE, "Partnerships for International Research and Education (PIRE)," webpage, undated.

42 NSF, *Accelerating Research Through International Network-to-Network Collaborations (AccelNet): Program Solicitation NSF 21-511*, undated.

43 NSF, OISE, "International Research Experiences for Students (IRES)," webpage, undated.

44 NSF, OISE, "NSF MULTIPLIER—MULTIPlying Impact Leveraging International Expertise in Research Missions," webpage, undated.

tional researchers are making significant developments and have the potential to benefit U.S. prosperity, security, health, and well-being.[45]

- NSF participates in international collaborations that enable large-scale international research facilities such as CERN and the International Gemini Observatory. The International Gemini Observatory is an international partnership among the United States, Canada, Chile, Brazil, Argentina, and the Republic of Korea that operates two 8.1-meter telescopes: one in the Northern Hemisphere on Maunakea, Hawaii, and one in the Southern Hemisphere on Cerro Pachón, Chile. The Association of Universities for Research in Astronomy, a consortium of 47 U.S. universities and institutions and three international affiliates, manages the Gemini Observatory for the international partners under a cooperative agreement with NSF. NSF acts as the executive agency for the partnership and provides the U.S. share of funding for the operation of the observatory.[46]

A.3.1.2. Permitting and Enabling the National Research Enterprise to Engage Internationally

Multiple NSF activities, while not dedicated to international cooperation, nonetheless are structured to allow it. For example,

- Researchers applying for NSF funds may include foreign collaborators and international travel and activities in their proposals (although NSF normally funds only the U.S. side).[47]
- National facilities that NSF manages through cooperative agreements to FFRDCs, often provide opportunities for foreign researchers to use their facilities. For example, the telescopes of the National Radio Astronomy Observatory "are open to all astronomers regardless of institutional or national affiliation."[48]
- Additionally, while NSF centers typically will not fund international partners, they often allow foreign entities to participate using separate funding. For example, the Centers for Chemical Innovation program will support costs for U.S. researchers to travel and work with foreign teams.[49] The Science and Technology Centers: Integrative Part-

[45] NSF, OISE, "NSF MULTIPLIER—MULTIPlying Impact Leveraging International Expertise in Research Missions," webpage, undated.

[46] NSF, "International Gemini Observatory," webpage, undated.

[47] NSF, *Proposal and Award Policies and Procedures Guide*, effective June 1, 2020, pp. I-6, II-10.

[48] NSF, National Radio Astronomy Observatory, "The NRAO Enables Forefront Research into the Invisible Universe," webpage, undated.

[49] NSF, *Centers for Chemical Innovation: Phase I Awards and New/Renewal Phase II Centers: Program Solicitation NSF 20-574*, undated.

nerships program encourages partnerships to include foreign organizations, but international participation is not required.[50]

A.3.1.3. Working with Partner Governments and International Organizations to Facilitate IBRC

NSF staff work with counterparts in foreign government and international organizations to establish joint funding agreements and arrangements with international partners when collaborations would provide a mutual scientific benefit. Such agreements allow research teams with U.S. and foreign collaborators to submit a single proposal and receive a single funding decision. For example, NSF and more than 20 foreign funding agencies established partnering arrangements to support the PIRE program.[51, 52] NSF interviewees noted the value of joint calls in avoiding the "double jeopardy" that occurs when research partners seek funding separately and run the risk of only a subset of the team being funded. Additionally, the leadership of NSF participates in international forums such as the Global Science Forum of the OECD, the International Group of Funding Agencies for Global Change Research, the activities of the Arctic Council, and the scientific activities of such United Nations specialized agencies as the World Meteorological Organization, among others.[53]

A.4. U.S. Department of Energy

DOE works to "advance U.S. national security and economic growth through transformative S&T innovation that promotes affordable and reliable energy through market solutions and meets our nuclear security and environmental cleanup challenges."[54] The DOE enterprise includes its headquarters in Washington, D.C., 83 field locations, and more than 100,000 federal and contractor employees.[55] DOE has 17 national laboratories that provide facilities used by researchers from the government, academia, and industry.[56] One, the National Energy Technology Laboratory, is an intramural (government-owned and government-operated)

[50] NSF, "Science and Technology Centers: Integrative Partnerships: Program Solicitation NSF 19-567," webpage, undated.

[51] NSF, OISE, "Partnerships for International Research and Education (PIRE)," webpage, undated.

[52] NSF, OISE, "Partnerships for International Research and Education (PIRE)," webpage, undated.

[53] NSF, OISE, "International Activities at NSF," webpage, undated.

[54] DOE, *Fiscal Year 2017 Annual Performance Report; Fiscal Year 2019 Annual Performance Plan*, undated, p. 4.

[55] DOE, *Fiscal Year 2017 Annual Performance Report; Fiscal Year 2019 Annual Performance Plan*, undated, p. 4.

[56] DOE, *Fiscal Year 2017 Annual Performance Report; Fiscal Year 2019 Annual Performance Plan*, undated, p. 4.

facility; the other 16 are FFRDCs.[57, 58] Although much of the work that DOE does is focused on applied science and engineering, DOE's Office of Science is the USG lead for supporting fundamental research on energy and production and is the top funder of physical science research in the United States.[59] Within DOE, the Office of International Affairs (IA) leads efforts to coordinate DOE's international cooperation.[60]

A.4.1. How the Department of Energy Facilitates and Supports International Basic Research Collaborations

DOE operates numerous international programs and maintains bilateral relationships that foster market development, commercialization, policy development and coordination, and collaborative research.[61, 62] Most of these efforts emphasize the later stages of technical and engineering development and focus on transitioning or commercializing scientific discoveries. However, some DOE efforts also include support for IBRC. Those efforts are summarized below using the categories described in Table 1.1.

A.4.1.1. Funding Research or Exploratory Activities with an International Component

DOE runs multiple activities specifically designed to foster international collaboration. For example,

- DOE supports international collaborations on large-scale scientific infrastructure, both within the United States and abroad. For example, DOE's Fermilab will host the Deep Underground Neutrino Experiment, which will combine the resources of more than 30

57 DOE, National Energy Technology Laboratory, "NETL Plays Key Role in DOE's New Lab Partnering Service Web Experience," webpage, July 11, 2018.

58 NSF, "Master Government List of Federally Funded R&D Centers," webpage, June 2020.

59 DOE, Office of Science, "About the Office of Science," webpage, undated.

60 DOE, Office of International Affairs, "Welcome to DOE International," webpage, April 18, 2014.

61 DOE, Office of Fossil Energy and Carbon Management, "Bilateral Agreements with China," webpage, undated.

62 DOE, Office of Nuclear Energy, "Bilateral Cooperation," webpage, undated.

countries, CERN, and 180 research institutions.[63, 64, 65, 66] Similarly, DOE has provided researchers and in-kind support to CERN.[67]

- DOE runs some joint funding calls with counterpart funding agencies, although one interviewee indicated that other mechanisms are preferred.

A.4.1.2. Permitting and Enabling the National Research Enterprise to Engage Internationally

Multiple DOE activities, while not dedicated to international cooperation, nonetheless allow it. For example,

- DOE national laboratory user facilities, which include high-performance computing capabilities, x-ray synchrotron radiation light sources, and high power lasers, are open to international researchers with the degree of access and the conditions depending on the facility. The DOE Office of Science's user facilities are "open to all interested potential users without regard to nationality or institutional affiliation."[68] Office of Science user facilities allocate access based on the merit of submitted proposals and do not charge for non-proprietary work that will be published in the open literature.[69] Nuclear Science User Facilities, which are managed by DOE's Office of Nuclear Energy, are more restrictive in that international researchers must have a U.S. collaborator to be eligible.[70, 71]
- U.S. national labs can enter into agreements to provide unique research and expertise to international and U.S. companies, nonprofit organizations, academia, and state or local governments.[72] Collaboration mechanisms include Cooperative Research and Develop-

[63] Fermilab, "DUNE at LBNF: Countries and Institutions Participating in DUNE," webpage, undated.

[64] Kurt Riesselmann, "UK Invests £65 Million in International Science Projects Hosted by Fermilab," Fermilab, press release, January 22, 2020.

[65] Leah Hesla, ed., "Fermilab, UNICAMP and São Paulo Research Foundation Collaborate on Major International Projects for Neutrino Research," webpage, Fermilab, March 13, 2020.

[66] Fermilab, "DUNE at LBNF: International Partnerships," webpage, undated.

[67] European Organization for Nuclear Research, "US Becomes Observer at CERN," press release, December 19, 1997.

[68] DOE, "Definition," webpage, undated.

[69] DOE, "Definition," webpage, undated.

[70] Nuclear Science User Facilities, "About Us," webpage, undated.

[71] Nuclear Science User Facilities, "Rapid Turnaround Experiments," webpage, undated.

[72] DOE, Office of Energy Efficiency & Renewable Energy, "How to Partner with the National Labs," undated.

ment Agreements, Strategic Partnership Projects, User Facility Agreements, technical service agreements, and licensing.[73]

- Some DOE funding opportunity announcements are open to non-domestic applicants. For example, non-domestic applicants who can demonstrate that they have capabilities and resources not available domestically are eligible for Office of Science's annual open solicitation.[74] Teams that include a mix of foreign and domestic partners are also eligible under that call.[75]

A.4.1.3. Working with Partner Governments and International Organizations to Facilitate IBRC

DOE engages with partner government and international organizations in multiple ways. (Most of the activities do not focus exclusively on basic research.) For example,[76]

- DOE stations some employees abroad, usually at U.S. embassies, who support all DOE activities, not just research.
- The IA office establishes a range of cooperation agreements with foreign government entities. Some agreements allow DOE labs to share and receive non-public data with foreign partners. Other agreements allow U.S. researchers to work in foreign facilities.
- DOE participates in international working groups and partnerships including the International Energy Agency, the U.S.-EU Energy Council, the Partnership for Transatlantic Energy Cooperation, the International Energy Forum, and many bilateral dialogs.[77]

---

[73] DOE, Office of Energy Efficiency & Renewable Energy, "How to Partner with the National Labs," undated.

[74] DOE, Office of Science, *FY 2020 Continuation of Solicitation for the Office of Science Financial Assistance Program, Funding Opportunity Announcement (FOA) Number: DE-FOA-0002181*, November 1, 2019.

[75] DOE, Office of Science, *FY 2020 Continuation of Solicitation for the Office of Science Financial Assistance Program, Funding Opportunity Announcement (FOA) Number: DE-FOA-0002181*, November 1, 2019.

[76] Unless indicated otherwise, the information came from one DOE interviewee.

[77] DOE, Office of International Affairs, "International Affairs Initiatives," webpage, undated.

APPENDIX B

Connections Between Security Cooperation and International Basic Research Collaboration at the U.S. Department of Defense

DoD distinguishes between security cooperation activities and other international efforts. According to the United States Code, security cooperation includes

> any program, activity (including an exercise), or interaction of the Department of Defense with the security establishment of a foreign country to achieve a purpose as follows:
>
> - To build and develop allied and friendly security capabilities for self-defense and multinational operations.
> - To provide the armed forces with access to the foreign country during peacetime or a contingency operation.
> - To build relationships that promote specific United States security interests.[1]

A complex set of laws and regulations, most of which are largely outside the scope of this project, govern DoD security cooperation activities. Of note, government-to-government R&D projects typically require international agreements and equitable contributions from all partners, which becomes relevant if basic research results transition to a government-government effort.[2] For example, a DoD interviewee explained that while international agreements are not needed to discuss basic research with foreign government security establishment counterparts, agreements are usually needed for discussions about applied research programs. The *Department of Defense International Science and Technology Engagement*

[1] U.S. Code, Title 10, Section 301, Definitions, January 3, 2012.

[2] See U.S. Code, Title 22, Section 2767, Arms Export Control Act: Authority of President to Enter into Cooperative Projects with Friendly Foreign Countries, 2012; U.S. Code, Title 10, Section 2350a, Cooperative Research and Development Agreements, January 7, 2011; U.S. Code, Title 10 USC, Section 2358, Research and Development Projects, January 5, 2015.

Strategy, which addresses all DoD international research cooperation (not just basic research), notes that agreements are foundational to the bulk of S&T cooperation.[3]

In contrast (per a DoD interviewee), research grants and contracts to foreign academic researchers are not international agreements, do not engage a foreign security establishment, and are not security assistance. Similarly, two DoD interviewees observed that the Terms of References under which AFOSR often coordinates grant funding with foreign partners are not international agreements, which allows them to be established comparatively quickly.

Although much of DoD's international basic research engagements do not require international agreements, two interviewees reported that when an agreement is needed it can take months to determine whether an applicable agreement already exists. If one does not, an interviewee indicated it can take even longer to establish one. One interviewee noted that the agreement process is largely "one-size-fits-all" with little consideration for the size of the investment or the associated risk. (DoD Instruction 5530.03, *International Agreements*, describes the processes DoD uses to negotiate and enter into international agreements.[4])

---

[3] DoD, *Department of Defense International Science and Technology Engagement Strategy: A Unified Approach to Strengthen Alliances and Attract New Partners*, October 2020.

[4] DoD, DoD Instruction 5530.03, *International Agreements*, December 4, 2019.

APPENDIX C

Overview of International Basic Research Collaboration Considerations for Microelectronics and Optoelectronics

To illustrate the IBRC considerations discussed in Chapter 4, this appendix provides an overview of two scientific fields, microelectronics and optoelectronics. The team chose microelectronics and optoelectronics as the fields for more in-depth analysis due to their current interest to the DoD BRO. Discoveries in both fields have supported a broad range of applications in the military and commercial sectors. While work in both fields is occurring across the TRL spectrum, this appendix focuses on basic research efforts.

Microelectronics research provides the foundation for the integrated circuits, logic devices, and memory components that are ubiquitous in modern society. For over five decades, numerous discoveries enabled the number of transistors in an integrated circuit to double approximately every two years resulting in the phenomenal performance gains and cost reductions predicted by Moore's law. As economic, physical, and materials limitations have ended Moore's law, the field is looking for new paradigms to fuel the next wave of advancements in computing. Basic research priorities include identifying new architectures, materials, and physical phenomena to exploit.[1]

Optoelectronics research examines ways to generate, modulate, detect, and switch photons in an analogous way to how electrical circuits use electrons.[2] Optoelectronics technologies are widely deployed in the commercial telecommunications sector, and other applications include chemical and biological sensing, night vision, and lasers.[3] While microelectronics has focused heavily on silicon-based materials, optoelectronics is exploring a broader range of

---

1 DOE, Office of Science, *Basic Research Needs for Microelectronics: Report of the Office of Science Workshop on Basic Research Needs for Microelectronics, October 3–25, 2018*, undated.

2 Peter Capper and Safa Kasap, eds., *Springer Handbook of Electronic and Photonic Materials*, Springer International Publishing, 2017.

3 NRC, *Optics and Photonics: Essential Technologies for Our Nation*, National Academies Press, 2013.

materials including III-V semiconductors (e.g., gallium arsenide, indium phosphide, gallium nitride), two dimensional materials (e.g., graphene), and perovskites.[4, 5, 6]

Terminology varies among sources. Semiconductors and microelectronics are often used interchangeably, particularly when referring to the associated industries. Additionally, optoelectronics are sometimes included as a sector within the overall microelectronics or semiconductor market.[7] Particularly in regards to basic research, most interviewees identified strong overlaps between the fields.

C.1. Collaboration Considerations

The remainder of this appendix looks at IBRC considerations in the two fields. First, it describes economic considerations and discusses the extent to which basic research and IBRC can impact the eventual commercial ecosystems. Then, it discusses considerations associated with the defense impacts and describes DoD's equities and recent laws and DoD programs intended to address those equities. The appendix next describes three considerations for obtaining scientific benefits in the fields through IBRC: (1) accessing unique and diverse resources, (2) combining resources, and (3) strengthening and building relationships. Next, it looks at how partner trustworthiness, the maturity of the fields, and the speed of transition from basic research to commercialization impact IBRC. The appendix ends by discussing overhead associated with IBRC for microelectronics and optoelectronics activities and the associated implications.

C.1.1. Economic Impacts

In 2019, the global semiconductor market generated $412.3 billion of sales, of which U.S.-headquartered firms accounted for 47 percent.[8] The United States has six of the largest 15 semiconductor firms and three of the five top equipment manufacturers.[9] The U.S. share of

---

4 Peter Capper and Safa Kasap, eds., *Springer Handbook of Electronic and Photonic Materials*, Springer International Publishing, 2017.

5 Fangxu Yang, Shanshan Cheng, Xiaotao Zhang, Xiaochen Ren, Rongjin Li, Huanli Dong, and Wenping Hu, "2D Organic Materials for Optoelectronic Applications," *Advanced Materials*, Vol. 30, No. 2, 2018.

6 Yani Chen, Yong Sun, Jiajun Peng, Junhui Tang, Kaibo Zheng, and Ziqi Liang, "2D Ruddlesden-Popper Perovskites for Optoelectronics," *Advanced Materials*, Vol. 30, No. 2, 2017.

7 For example, Platzer, Sargent, and Sutter, 2020, identified "optoelectronics, sensors, and discretes" as one segment of the semiconductor market. See Michaela D. Platzer, John F. Sargent, Jr., and Karen M. Sutter, *Semiconductors: U.S. Industry, Global Competition, and Federal Policy*, Congressional Research Service, R46581, 2020, p. 8.

8 Semiconductor Industry Association, *2020: State of the U.S. Semiconductor Industry*, undated.

9 Michaela D. Platzer, John F. Sargent, Jr., and Karen M. Sutter, *Semiconductors: U.S. Industry, Global Competition, and Federal Policy*, Congressional Research Service, R46581, 2020, p. 1.

global semiconductor manufacturing capacity, however, has declined from 40 percent in 1990 to 11 percent in 2019.[10] This trend is largely rooted in U.S. firms' shift to a fabless manufacturing model, in which manufacturing is contracted to external providers. Indeed, of the six large semiconductor firms based in the United States, three are fabless (seven of the ten largest fabless firms are located within the United States).[11] This decline in the share of global manufacturing capacity has coincided with a decline in the technical ability needed to manufacture state-of-the-art technologies,[12] a growth in reliance on foreign suppliers,[13] an increased emphasis among adversarial governments on domestic semiconductor manufacturing,[14] and immediate-term supply shortages,[15] raising concerns about the U.S. commercial sector's and DoD's short-term and long-term access to semiconductors. (See the next section for more on defense considerations.)

Table C.1 shows factors that semiconductor firms examine when selecting manufacturing locations, their relative importance, and the relative rank of the United States compared with other nations. Basic research and IBRC support some, but not all, of the factors. For example, basic research and IBRC can contribute to a synergistic ecosystem of government, academic, and commercial stakeholders that facilitates technology transfer and provides access to facilities and ideas, which is the most important factor.[16] Basic research and IBRC also contribute to the development of the talent pool accessible to commercial firms, particularly in the high-skill, high-wage positions that semiconductor manufacturing requires.[17] As economical manufacturing is a microelectronics basic research priority topic, basic research and IBRC might also assist with lowering required capital expenditures.[18] The other factors, however, depend on governmental policies and national systems that produce favorable economic cli-

[10] Michaela D. Platzer, John F. Sargent, Jr., and Karen M. Sutter, *Semiconductors: U.S. Industry, Global Competition, and Federal Policy*, Congressional Research Service, R46581, 2020, p. 1.

[11] Michaela D. Platzer, John F. Sargent, Jr., and Karen M. Sutter, *Semiconductors: U.S. Industry, Global Competition, and Federal Policy*, Congressional Research Service, R46581, 2020, p. 1.

[12] Will Knight, "Intel Wants to Revive U.S. Chipmaking—But It Has to Catch Up First," *Wired*, March 24, 2021.

[13] Kevin Curran, "Is Taiwan a Ticking Time Bomb in the Semiconductor Supply Chain?" *The Diplomat*, February 27, 2021.

[14] Justin Hodiak and Scott W. Harold, "Can China Become the World Leader in Semiconductors?" *The Diplomat*, September 25, 2020.

[15] Sam Shead, "The Global Chip Shortage Is Starting to Have Major Real-World Consequences," *CNBC*, May 7, 2021.

[16] European Commission DG Communications Networks, Content & Technology, "Comparison of European and Non-European Regional Clusters in KETs [Key Enabling Technologies]: The Case of Semiconductors," webpage, 2013.

[17] Michaela D. Platzer, John F. Sargent, Jr., and Karen M. Sutter, *Semiconductors: U.S. Industry, Global Competition, and Federal Policy*, Congressional Research Service, R46581, 2020.

[18] DOE, Office of Science, *Basic Research Needs for Microelectronics: Report of the Office of Science Workshop on Basic Research Needs for Microelectronics, October 3–25, 2018*, undated.

TABLE C.1
Factors That Semiconductor Firms Use to Select Manufacturing Locations

Importance (Rank)	Factor	U.S. Above Median?	Basic Research Supports Factor?	IBRC Supports Factor?
1	Synergies with ecosystem/ footprint	Yes	Yes	Yes
2	Labor costs			
2	Access to talent	Yes	Yes	Yes
4	Government incentives			
4	Security of IP/assets	Yes		
6	Capital expenditure		Possibly	Possibly
7	Supporting infrastructure			
7	Ease of doing business	Yes		
9	Geopolitical considerations	Yes		

SOURCE: Table was adapted from Antonio Varas, Raj Varadarajan, Jimmy Goodrich, and Falan Yinug, "Government Incentives and U.S. Competitiveness in Semiconductor Manufacturing," *BCG and Semiconductor Industry Association*, September 2020, p. 15, with RAND analysis on the influence of basic research and IBRC.

mates and regulatory regimes and are largely independent of basic research. Favorable policies can include, for example, tax breaks; subsidies; and access to land, electricity, and water.[19] Multiple interviewees explained that in view of the long time horizon of basic research, other government policies (e.g., tax incentives for companies) would be more effective than basic research investments in building near-term manufacturing capacity.

A 2013 National Research Council report titled *Optics and Photonics: Essential Technologies for Our Nation* examined the relationships between the locations of manufacturing and innovation for three case studies.[20] The report concluded that the details of the cases and the relative tightness between R&D and manufacturing in the subfields for each case impacted the outcomes and that government policies to incentivize manufacturing in the United States should be tailored to specific specialties rather than uniform for all of photonics.[21]

While the relationship between the location of discoveries and their commercialization is complex, multiple DoD interviewees emphasized the need to consider transition options in these fields even at the basic research stage. Several indicated that IBRC could in some cases

---

19 Antonio Varas, Raj Varadarajan, Jimmy Goodrich, and Falan Yinug, "Government Incentives and U.S. Competitiveness in Semiconductor Manufacturing," *BCG and Semiconductor Industry Association*, September 2020.

20 NRC, *Optics and Photonics: Essential Technologies for Our Nation*, National Academies Press, 2013, pp. 186–202.

21 NRC, *Optics and Photonics: Essential Technologies for Our Nation*, National Academies Press, 2013, p. 202.

be used to gain a competitive advantage, and that some technologies developed by foreign DoD performers have transitioned to U.S. industry. Interviewees also observed that technology ecosystems can grow with or without the United States and that the risks of not being engaged should not be underestimated.

C.1.2. Defense Impacts

DoD has a variety of semiconductor needs including state-of-the-art, commercial microelectronics for cutting edge applications; microchips with classified or sensitive designs; legacy integrated circuits no longer produced by commercial production lines to sustain systems; radiation hardened microelectronics; and optoelectronics for sensors.[22, 23] While some of DoD's needs are unique, its ability to influence the semiconductor market has declined as its share of global semiconductor demand has shrunk. In the 1970s, 7 percent of the integrated circuit market supported defense applications.[24] By 2005, the defense share had dropped to 1 to 2 percent.[25] By 2019, government (not just defense) users accounted for just 1.3 percent of semiconductor sales.[26] DoD runs multiple programs to address its semiconductor needs (Table C.2). Additionally, interviewees indicated that the DoD service laboratories sponsor a variety of microelectronics and optoelectronics research that strives to advance the state of the art, help DoD maintain situational awareness of emerging breakthroughs, and promote accessible technology transition pathways.

Several recent laws and executive orders are intended to help DoD and the United States meet its semiconductor needs. Section 276 of the FY 2021 National Defense Authorization Act (NDAA) requires DoD to assess microelectronics supply chain management risks and best practices; develop a plan to increase the commercialization of DoD's microelectronics IP; assess options for public-private partnerships; evaluate the utility of creating a national laboratory focused on microelectronics R&D; and develop criteria and processes for selecting commercial microelectronics suppliers, including companies whose headquarters are located in allied or partner countries.[27] Sections 9901–9908 of the FY 2021 NDAA, Creating Helpful Incentives to Produce Semiconductors (CHIPS) for America Act, authorizes multiple initiatives to advance the domestic semiconductor technology ecosystem, such as a mechanism

---

22 DSB, *High Performance Microchip Supply*, Office of the Under Secretary of Defense for Acquisition, Technology, and Logistics, February 2005.

23 Saif M. Khan and Alexander Mann, "AI Chips: What They Are and Why They Matter," Center for Security and Emerging Technology, April 2020.

24 DSB, *High Performance Microchip Supply*, Office of the Under Secretary of Defense for Acquisition, Technology, and Logistics, February 2005, p. 17.

25 DSB, *High Performance Microchip Supply*, Office of the Under Secretary of Defense for Acquisition, Technology, and Logistics, February 2005, p. 17.

26 Semiconductor Industry Association, *2020: State of the U.S. Semiconductor Industry*, undated.

27 National Defense Authorization Act for FY 2021, Section 276, Microelectronics and National Security.

TABLE C.2
Selected Department of Defense Microelectronics and Optoelectronics Priorities and Associated Programs

Priority	Associated Program (Responsible Organization)
• Access state-of-the-art commercial technology	• Trusted Supplier Program (DMEA) • Trusted Access Program Office (DMEA)
• Ensure microelectronic component integrity through data driven quantifiable assurance	• Joint Federated Assurance Center (OSD)
• Address DoD unique needs (e.g., radiation-hardened electronics, specialized radio frequencies and electro-optic components)	• Trusted Foundry Program (DMEA)
• Create resilient and robust technology and talent pipelines	• Trusted and Assured Microelectronics (OSD) • Microelectronics Innovation for National Security and Economic Competitiveness (OSD) • Electronic Resurgence Initiative (DARPA) • Trusted Integrated Chips (IARPA)

SOURCES: Matthew Casto, "Trusted and Assured Microelectronics Program," DARPA ERI Summit, August 2020; Jeremy Muldavin, "DoD Trusted and Assured Microelectronics Summary," NDIA Electronics Division Meeting, February 7, 2019; and Carl E. McCants, "IARPA Trusted Integrated Chips (TIC) Program," Office of the Director of National Intelligence, August 17, 2016.

NOTE: DMEA = Defense Microelectronics Activity; IARPA = Intelligence Advanced Research Projects Activity.

to co-fund secure semiconductors and semiconductor supply chains with foreign partners, including enabling R&D.[28] (As of July 2021, funding for these efforts had not been appropriated.) Furthermore, on February 24, 2021, President Biden signed the executive order "Securing America's Critical Supply Chains," which requires federal agencies to review and address vulnerabilities in semiconductor and advanced packaging supply chains.[29] The aforementioned initiatives notwithstanding, multiple DoD interviewees indicated that additional strategic planning to align national, DoD, and service research investments with strategic objectives would be useful.

As with economic considerations, defense considerations cause DoD to think carefully about transition options for IBRC projects.

The proximity of many basic research fields within microelectronics and optoelectronics to applied fields that are restricted by ITAR also challenges IBRC because with slight modification many basic research topics can become applied and subject to ITAR. (Additionally, ITAR restrictions on specific sub-disciplines such as some types of high-power lasers can impact basic research directly.) An interviewee indicated that concerns about what could be inferred from non-ITAR-controlled results have stopped DoD from allowing some papers to be published. Multiple DoD researchers indicated that they worked on two separate types of

[28] National Defense Authorization Act for FY 2021, Sections 9901–9908, Creating Helpful Incentives to Produce Semiconductors (CHIPS) for America Act.

[29] White House, "Fact Sheet: Securing America's Critical Supply Chains," February 24, 2021.

projects: basic research projects in which they were encouraged to find the best partners—foreign or domestic—and sensitive or classified research in which they excluded foreign companies and universities with foreign graduate students from the work.

Interviewees also discussed the importance of keeping restrictions in the right places as technologies advance and noted that being overly cautious can be counterproductive. For example, a DoD interviewee described how the United States lost leadership in one optoelectronics application area after classifying most work on the topic; the Europeans and Canadians kept working openly and ultimately replaced the United States as the world leaders. Similarly, some companies have moved manufacturing and supporting R&D overseas to avoid having commercial applications become subject to ITAR controls.[30]

C.1.3. Ability to Combine Resources and Share the Burden

The microelectronics sector is capital intensive, with leading edge manufacturing facilities costing upwards of $20 billion and critical equipment exceeding $100 million.[31] Optoelectronics and microelectronics research often requires access to advanced facilities that could exceed the budget constraints of individual institutions or countries. (An interviewee noted that the optoelectronics field is easier to enter than microelectronics because large investments are less necessary.) Specialized research infrastructure includes synchrotron facilities (e.g., Argonne National Laboratory's Advanced Photon Source, the European Synchrotron Radiation Facility), advanced manufacturing capabilities (e.g., New York-based AIM Photonics, Belgium's Interuniversity Microelectronics Centre), and clean room space (e.g., SUNY Polytechnic Institute's NanoTech Complex, the Republic of Korea's National Nano Fab Center). An academic interviewee indicated that research infrastructure typically needs regular, costly upgrades and maintenance to remain cutting edge. The interviewee elaborated that high research infrastructure costs combined with the global nature of the semiconductor value chain has resulted in a distributed web of expertise and state-of-the-art facilities. Finite budgets and institutional capacity often results in institutions and countries specializing in subfields, such as advanced packaging or integrated photonics. This generates opportunities for partners to combine resources and (as discussed in the next section) access unique resources.

Within the United States, public-private partnerships tend to span TRLs, be U.S.-centric, and primarily involve U.S.-based researchers. Although foreign-headquartered semiconductor firms (e.g., Samsung, Taiwan Semiconductor Manufacturing Company) have engaged in some of these efforts, such as the Global 450 Consortium at SUNY Polytechnic University,[32]

---

30 NRC, *Optics and Photonics: Essential Technologies for Our Nation*, National Academies Press, 2013, pp. 121–123.

31 Michaela D. Platzer, John F. Sargent, Jr., and Karen M. Sutter, *Semiconductors: U.S. Industry, Global Competition, and Federal Policy*, Congressional Research Service, R46581, 2020.

32 Larry Rulison, "NY: Demise of G450C Wasn't over Money," *Times Union*, January 13, 2017.

DoD-funded public-private partnerships (e.g., SEMATECH, Nanoelectronics for 2020 and Beyond, the Electronics Resurgence Initiative) tend to exclude foreign companies.[33] In contrast, Europe is increasingly prioritizing transnational public-private partnerships that are jointly funded by industry, research organizations, participating EU-member states, and the European Commission.[34, 35] While assessments of both the U.S. domestic-centric partnership model and European transnational partnerships indicate that positive technological and economic benefits were achieved,[36, 37] the relative benefits of the two models (international versus domestic) are unknown.

One aspect of pooling resources is leveraging the investments of others and allowing them to leverage yours. One interviewee indicated that funding foreign performers can allow DoD to use its research funds more efficiently, while acknowledging that working abroad does not have the benefit of training the U.S. workforce. Investing in U.S. infrastructure attracts good researchers from throughout the world. While some foreign researchers bring their expertise and results back to their home countries, DoD interviewees indicated that the alternative is to reduce the rate of progress and to try to compete blind.

Some interviewees emphasized the importance of the United States, including the DoD intramural laboratories, having indigenous capabilities. They argued that buying cheaper goods and services from abroad was a false economy and created unacceptable vulnerabilities. Perspectives varied over the extent to which the United States should pool resources with partners to provide critical capabilities and with which partners the United States should work.

C.1.4. Access to Unique and Diverse Resources

As noted above, basic microelectronics and optoelectronics research requires cutting-edge facilities to support advanced computational modeling, materials synthesis, and characterization techniques. Advancements within these fields also draw on expertise in wide-ranging

---

[33] National Nanotechnology Initiative, "Highlights from the Nanoelectronics for 2020 and Beyond (Nanoelectronics) NSI [Nanotechnology Signature Initiative]," April 2017.

[34] Haydn Thompson, Emilio Lora-Tamayo, Damm Werner, Jean-Luc Dormoy, Leonard Hobbs, Margriet Jansz, Tomasz Kosmider, and Wolfgang Pribyl, "Final Evaluation of the ARTEMIS and ENIAC Joint Undertaking (2008-2013) Operating Under FP7, Directorate-General for Communications Networks, Content and Technology (European Commission)," June 2017.

[35] Silicon Europe, "Silicon Europe: Cluster Alliance for European Micro- and Nanoelectronics Industry," press release, October 8, 2012.

[36] Haydn Thompson, Emilio Lora-Tamayo, Damm Werner, Jean-Luc Dormoy, Leonard Hobbs, Margriet Jansz, Tomasz Kosmider, and Wolfgang Pribyl, "Final Evaluation of the ARTEMIS and ENIAC Joint Undertaking (2008-2013) Operating Under FP7, Directorate-General for Communications Networks, Content and Technology (European Commission)," June 2017.

[37] U.S. General Accounting Office, "Lessons Learned from SEMATECH," GAO/RCED-92-283, September 1992.

fields, from condensed matter physics to computational modeling to algorithm design.[38] Multiple interviewees indicated that microelectronics and optoelectronics basic research is distributed among multiple countries, so engaging and collaborating with foreign researchers can provide access to unique ideas, research infrastructure, and techniques.

DoD interviewees identified multiple ways that DoD uses IBRC to access diverse resources. Attending international conferences is an opportunity to exchange ideas and see what tools different researchers could contribute, especially those more advanced than what are currently available in the United States. One DoD IOP gave an example of a country with few facilities but excellent theorists whose work complemented that of DoD intramural researchers who rarely had time for theory work. Others noted the value in diversity of thought, as well as the more tangible value in collaborating with good people who can do specialized techniques or work with unique materials (e.g., gallium antimonide). DoD also participates in Cooperative Research and Development Agreements with foreign companies and universities.

C.1.5. Build and Strengthen Relationships

As with DoD IBRC in other fields, DoD interviewees pointed to the value that DoD gains from forming relationships with international researchers and connecting to their scientific networks and gaining awareness. Research collaborations and the relationships they create can positively impact technology transitions into domestic technology ecosystems, especially when the commercial sector is engaging with researchers in public-private partnerships. For example, SEMATECH, a public private partnership between researchers and 14 U.S. semiconductor firms that DARPA funded from FY 1988 to FY 1996, contributed to the reversal of historically declining U.S. semiconductor manufacturing capacity.[39] Ongoing collaborations in Europe (which tend to be transnational), the United States, China, and Taiwan have sustained innovation clusters that enable technology maturation and commercial transitions.[40]

C.1.6. Trust in a Partner

Due to the large defense and economic impacts of optoelectronics and microelectronics advances, the consideration of how much trust should be placed in a potential partner is also highly relevant. (While public information on this consideration for microelectronics is abundant, information specific to optoelectronics is not.)

---

38 NRC, *Materials Research to Meet 21st-Century Defense Needs*, National Academies Press, 2003.

39 Michaela D. Platzer, John F. Sargent, Jr., and Karen M. Sutter, *Semiconductors: U.S. Industry, Global Competition, and Federal Policy*, Congressional Research Service, R46581, 2020.

40 European Commission DG Communications Networks, Content & Technology, "Comparison of European and Non-European Regional Clusters in KETs [Key Enabling Technologies]: The Case of Semiconductors," webpage, 2013.

Foreign militaries, intelligence communities, and defense industrial bases are increasingly deploying illicit mechanisms to access IP.[41] Microelectronics has been an attractive area for IP theft and illicit engagement with U.S. scientists, indicating the intent of some potential partners to misuse research results.[42, 43]

Additionally, other nations, including U.S. competitors, are investing significantly in microelectronics through direct state funding, tax incentives, and public-private partnerships.[44] Thus, multiple countries are becoming increasingly able to apply microelectronics information—regardless of whether that information is acquired legitimately or illicitly.

To address trust considerations, the CHIPS Act for America prohibits funding "foreign entities of concern" and directs that potential partners be screened for influence by such entities.[45] Some interviewees, however, questioned the practicality of examining partners of their partners and wondered about where to bound the inquiries and what types of interactions were acceptable at what distance in a network. The export restrictions and the associated cautions discussed in the defense impacts section are also relevant for this consideration. Academic and DoD interviewees also indicated that with any partner, domestic or international, it is prudent to consider carefully what to share and what to withhold.

C.1.7. Maturity and Speed of Transition Within a Scientific Field

Although there is a significant amount of overlap between microelectronics and optoelectronics, interviewees indicated that the optoelectronics field is generally less mature than microelectronics. An interviewee indicated that optoelectronics researchers are experimenting with a range of materials and methods, with foreign researchers occupying some promising niches. DoD IOP interviewees also indicated that the small grants they provide to foreign researchers are an economical way to explore a range of approaches.

Because aspects of optoelectronics are less mature than microelectronics and the applications are still being discovered, an interviewee indicated that there might still be time to prevent the offshoring of some areas of optoelectronics. Another interviewee suggested that the best way for the United States to influence the fate of optoelectronics would be continue pushing the field forward, as the country in the lead tends to have the most control. However,

41 Michael Brown and Pavneet Singh, *China's Technology Transfer Strategy: How Chinese Investments in Emerging Technology Enable a Strategic Competitor to Access the Crown Jewels of U.S. Innovation*, Defense Innovation Unit Experimental, January 2018.

42 Debby Wu, "Engineers Found Guilty of Stealing Micron Secrets for China," *Bloomberg*, June 12, 2020.

43 Veronica Stracqualursi and Sheena Jones, "Harvard Professor Among Three Charged with Lying About Chinese Government Ties," *CNN*, January 28, 2020.

44 European Commission DG Communications Networks, Content & Technology, "Comparison of European and Non-European Regional Clusters in KETs [Key Enabling Technologies]: The Case of Semiconductors," webpage, 2013.

45 National Defense Authorization Act for FY 2021, Sections 9901–9908, Creating Helpful Incentives to Produce Semiconductors (CHIPS) for America Act.

the nonresearch factors discussed in the economic impacts section are also relevant, and interviewees indicated that technology transitions from foreign countries to U.S. industry are often challenging to facilitate.

C.1.8. Amount of Collaboration-Related Overhead

The DoD personnel interviewed about microelectronics and optoelectronics identified many of the common concerns DoD intramural researchers have about the time and effort needed to identify, evaluate, initiate, and conduct IBRCs (see Section 5.2.3). Of note, time needed to obtain approvals for collaborations was a substantial concern and deterrent. The effort required to ensure that basic research projects with foreign partners did not evolve into ITAR-controlled applied areas was also a concern.

APPENDIX D

Challenges Associated with International Research Collaborations and Contributors to and Inhibitors of Success

While scholars in the field of research collaborations consider international collaborations to be a subset of collaborations, international collaborations do have unique challenges.[1] Table 5.2 summarized contributors to successful international research collaborations (most are not unique to basic research) along with the associated inhibitors. Each is discussed in more detail below.

D.1. Establish the Terms of Engagement Early

Ensuring that all parties have a shared understanding of a collaboration is one success factor, and planning conversations and agreements can help set the stage for a successful collaboration.[2] Early conversations help participants understand their commonalities and differences, allowing them to address any differing expectations about goals, resource contributions, information sharing, or activities.[3] Some suggest establishing processes for handling disputes and completing a formal MOU.[4] Others, however, note the value of the speed and flexibility more informal governance can provide.[5]

---

[1] J. Sylvan Katz and Ben Martin, "What Is Research Collaboration?" *Research Policy*, Vol. 26, No. 1, 1997.

[2] Elisabeth Pain, "Tips for Successful International Collaborations," *Science*, February 15, 2013.

[3] National Academy of Sciences, National Academy of Engineering, and Institute of Medicine, *Examining Core Elements of International Research Collaboration: Summary of a Workshop*, National Academies Press, 2011.

[4] National Academy of Sciences, National Academy of Engineering, and Institute of Medicine, *Examining Core Elements of International Research Collaboration: Summary of a Workshop*, National Academies Press, 2011.

[5] Aurelia Stamm and Andreas Figueroa, "Effective International Science, Technology and Innovation Collaboration: From Lessons Learned to Policy Change," in OECD, *Meeting Global Challenges Through Better Governance: International Co-operation in Science, Technology and Innovation*, OECD Publishing, 2012.

D.2. Free, Open, and Clear Communication with Cultural Understanding

Communication is essential for all collaborations, but it sometimes is more complicated when participants are in different countries. Not understanding a partner's culture can result in miscommunications and misunderstandings of research protocols and the interpretation of research information.[6, 7] In successful collaborations, communication follows principles of trust, transparency, and openness.[8, 9, 10] If language barriers exist, securing translators can facilitate communication.[11] Collaborations where efforts have been undertaken to understand and respect cultural differences and build partnerships tend to proceed more successfully.[12]

D.3. Effective Leadership

Multiple scholars argue that effective leadership is an important factor that contributes to successful collaborations in general but also specifically for international research collaborations.[13, 14, 15] Leadership is needed to guide all parties as they perform their specific activities, encounter problems, and face questions regarding resources including those of funding and division of labor. Ineffective leadership contributed to the cost overruns and schedule delays

---

6 National Academy of Sciences, National Academy of Engineering, and Institute of Medicine, *Examining Core Elements of International Research Collaboration: Summary of a Workshop*, National Academies Press, 2011.

7 Dawn Freshwater, Gwen Sherwood, and Vicki Drury, "International Research Collaboration: Issues, Benefits and Challenges of the Global Network," *Journal of Research in Nursing*, July 1, 2006.

8 Elisabeth Pain, "Tips for Successful International Collaborations," *Science*, February 15, 2013.

9 National Academy of Sciences, National Academy of Engineering, and Institute of Medicine, *Examining Core Elements of International Research Collaboration: Summary of a Workshop*, National Academies Press, 2011.

10 Derek Byerlee and H. J. Dubin, "Crop Improvement in the CGIAR as a Global Success Story of Open Access and International Collaboration," *International Journal of the Commons*, Vol. 4, No. 1, 2010.

11 Dawn Freshwater, Gwen Sherwood, and Vicki Drury, "International Research Collaboration: Issues, Benefits and Challenges of the Global Network," *Journal of Research in Nursing*, July 1, 2006.

12 Dawn Freshwater, Gwen Sherwood, and Vicki Drury, "International Research Collaboration: Issues, Benefits and Challenges of the Global Network," *Journal of Research in Nursing*, July 1, 2006.

13 National Academy of Sciences, National Academy of Engineering, and Institute of Medicine, *Examining Core Elements of International Research Collaboration: Summary of a Workshop*, National Academies Press, 2011.

14 Derek Byerlee and H. J. Dubin, "Crop Improvement in the CGIAR as a Global Success Story of Open Access and International Collaboration," *International Journal of the Commons*, Vol. 4, No. 1, 2010.

15 Barry Bozeman and Jan Youtie, *The Strength in Numbers: The New Science of Team Science*, Princeton University Press, 2017.

that the International Thermonuclear Experimental Reactor, the international fusion reactor project in France, experienced.[16]

D.4. Diversity of Participants

Many studies point to diversity as a reason to collaborate as it allows for wider perspectives and approaches to research, which are associated with innovation[17] and more replicable findings.[18] Collaborations benefit from attracting researchers with complementary skills, backgrounds, and training.[19]

D.5. Sufficient and Flexible Funding

Funding is important for all research projects. Without sufficient and sustainable funding, a collaboration is unlikely to obtain critical resources, and thus has a higher likelihood of failing.[20, 21, 22] International collaborations, in particular, benefit from flexible funding, such as funds that can be used outside of a researcher's country.[23] Additionally, aligning national and international science programs can help justify collaborations and secure funds.[24]

---

16 Daniel Clery, "New Review Slams Fusion Project's Management," *Science*, Vol. 343, No. 6174, February 28, 2014.

17 Alina Lungeanu and Noshir S. Contractor, "The Effects of Diversity and Network Ties on Innovations: The Emergence of a New Scientific Field," *American Behavioral Scientist*, Vol. 59, No. 5, May 1, 2015.

18 Valentine Danchev, Andrey Rzhetsky, and James A. Evans, "Meta-Research: Centralized Scientific Communities Are Less Likely to Generate Replicable Results," *Elife*, Vol. 8, e43094, July 2, 2019.

19 Sooho Lee and Barry Bozeman, "The Impact of Research Collaboration on Scientific Productivity," *Social Studies of Science*, Vol. 35, No. 5, October 1, 2005.

20 Brian Owens, "The Benefits and Challenges of International Research Collaboration," *University Affairs*, November 7, 2018.

21 Derek Byerlee and H. J. Dubin, "Crop Improvement in the CGIAR as a Global Success Story of Open Access and International Collaboration," *International Journal of the Commons*, Vol. 4, No. 1, 2010.

22 Michael J. Adams, "Mutual Benefit: Building a Successful Collaboration," *Science*, October 6, 2000.

23 Baker Institute for Public Policy, Rice University, *Science Collaboration Across Borders*, Baker Institute Policy Report, No. 42, August 2009.

24 Aurelia Stamm and Andreas Figueroa, "Effective International Science, Technology and Innovation Collaboration: From Lessons Learned to Policy Change," in OECD, *Meeting Global Challenges Through Better Governance: International Co-operation in Science, Technology and Innovation*, OECD Publishing, 2012.

D.6. Knowledge of Relevant Regulations and Processes

In any collaboration, bureaucracies can be a barrier to progress, and international collaborations require participants to invest the time and resources to navigate multiple bureaucracies. Ignorance of bureaucratic processes in another country can lead to delays in starting the collaboration, issues with exchanging information, delays in obtaining visas, and in some cases corruption.[25]

D.7. Regulations That Facilitate Researcher Mobility and Information Exchange

The regulatory environment itself can also impact collaboration effectiveness, ease, and success.[26, 27] Some scholars suggest that collaborations are most successful where there are scientific visa provisions to allow for easier mobility for researchers.[28] Tailored IP agreements acceptable to the participants and research sponsors can also be helpful.[29]

---

25 Elisabeth Pain, "Tips for Successful International Collaborations," *Science*, February 15, 2013.

26 Terhi Nokkala, *Internal Collaboration Rules in International R&D Collaboration Projects—Analysis of Seven NEST Projects*, NEMO, NEMO Working Paper #15, January 2009.

27 J. L. Culhane and J. C. Worms, "Improving the Effectiveness of International Collaboration in Space Science," *Space Policy*, Vol. 17, 2001.

28 Stefano Lami, "Challenges and New Requirements for International Collaborations," *Science & Diplomacy*, Vol. 6, No. 2, June 2017.

29 Aurelia Stamm and Andreas Figueroa, "Effective International Science, Technology and Innovation Collaboration: From Lessons Learned to Policy Change," in OECD, *Meeting Global Challenges Through Better Governance: International Co-operation in Science, Technology and Innovation*, OECD Publishing, 2012.

Abbreviations

AccelNet	Accelerating Research Through International Network-to-Network Collaborations
AFOSR	Air Force Office of Scientific Research
AFOSR/IO	Air Force Office of Scientific Research International Office
AFRL	U.S. Air Force Research Laboratory
AI	artificial intelligence
ANR	Agence Nationale de la Recherche [National Research Agency of France]
ARO	U.S. Army Research Office
ASTRID	Accompagnement Spécifique des Travaux de Recherches et d'Innovation Défense (Specific Support for Defense Research and Innovation)
BRO	Basic Research Office
CAS	Chinese Academy of Sciences
CCDC	U.S. Army Combat Capabilities Development Command
CERN	European Organization for Nuclear Research
CNRS	Centre National de la Recherche Scientifique (National Scientific Research Center)
COVID-19	coronavirus disease 2019
DARPA	Defense Advanced Research Projects Agency
DEPSCoR	Defense Established Program to Stimulate Competitive Research
DGA	Direction Générale de l'Armement (Directorate General of Armaments)
DoD	U.S. Department of Defense
DOE	U.S. Department of Energy
ELI	Extreme Light Infrastructure
ERA	European Research Area
ESEP	Engineer and Scientist Exchange Program

EU	European Union
FFRDC	federally funded research and development center
FY	fiscal year
HHS	U.S. Department of Health and Human Services
IBRC	international basic research collaboration
ICE-PPR	International Cooperative Engagement Program for Polar Research
ICT	information and communications technology
IOP	international office person
IP	intellectual property
ITAR	International Traffic in Arms Regulations
ISEP	International Supplemental Student Exchange Program
MOST	Chinese Ministry of Science and Technology
MOU	memorandum of understanding
MULTIPLIER	MULTIPlying Impact Leveraging International Expertise in Research Missions
NAE	National Academy of Engineering
NASA	National Aeronautics and Space Administration
NATO	North Atlantic Treaty Organization
NRC	National Research Council
NSB	National Science Board
NSDD	National Security Decision Directive
NSF	National Science Foundation
NSFC	National Natural Science Foundation of China
NSTC	National Science and Technology Council
OECD	Organisation for Economic Co-operation and Development

OISE	NSF Office of International Science and Engineering
ONR	Office of Naval Research
OSD	Office of the Secretary of Defense
OSTP	Office of Science and Technology Policy
OUSD(R&E)	Office of the Under Secretary of Defense for Research and Engineering
PI	principal investigator
PIRE	Partnerships for International Research and Education
PPP	purchasing power parity
PRC	People's Republic of China
PRCI	projets de recherche collaborative—international [international collaborative research projects]
R&D	research and development
S&T	science and technology
SESAME	Synchrotron-light for Experimental Science and Applications in the Middle East
SME	subject-matter expert
TRL	technology readiness level
UK	United Kingdom
USDA	U.S. Department of Agriculture
USG	U.S. government
USSR	Union of Soviet Socialist Republics

References

Abbott, Alison, "World's Largest Laser Lab Rocked by Slew of Disputes," *Nature*, No. 569, May 23, 2019 (corrected May 28, 2019), pp. 607–608.

Acharya, Ram C., and Wolfgang Keller, *Technology Transfer Through Imports*, NBER Working Paper No. 13086, May 2007.

Adams, Jonathan, Karen Gurney, and Stuart Marshall, "Patterns of International Collaboration for the UK and Leading Partners (Summary Report)," *Evidence*, 2007.

Adams, Michael J., "Mutual Benefit: Building a Successful Collaboration," *Science*, October 6, 2000.

AFOSR—*See* Air Force Office of Scientific Research.

AFRL—*See* U.S. Air Force Research Laboratory.

Agence Nationale de la Recherche, "Missions," webpage, undated. As of February 1, 2021: https://anr.fr/en/anrs-role-in-research/missions/

Agence Nationale de la Recherche, *Soutenir la recherche dans sa diversité—Funding Research in All Its Diversity: Rapport d'Activité 2019—Annual Report 2019*, undated.

Agence Nationale de la Recherche, "Transnational Collaborations," webpage, undated. As of February 2, 2021:
https://anr.fr/en/anrs-role-in-research/europe-and-international/transnational-collaborations/

Agence Nationale de la Recherche, "Transnational Cooperation Strategy," webpage, undated. As of February 1, 2021:
https://anr.fr/en/anrs-role-in-research/europe-and-international/transnational-cooperation-strategy/

Agence Nationale de la Recherche, *Reglement Financier Relatif aux Modalités d'Attribution des Aides de l'ANR* [*Financial Regulations on the Terms of Allocation Aid from the ANR*], July 2019.

Agence Nationale de la Recherche, *Accompagnement Spécifique des Travaux de Recherches et d'Innovation Défense: ASTRID* [*Specific Support for Defense Research and Innovation Work: ASTRID*], 2020.

Agence Nationale de la Recherche, *2021 Work Programme: Generic Call for Proposals: 2021 AAPG Guide: Submission, Evaluation, Selection and Funding Guidelines*, English version, v.1.0, October 2, 2020.

Agence Nationale de la Recherche, *Work Programme 2021*, English version, v.1.1, October 22, 2020.

Air Force Office of Scientific Research, "MOST-AFOSR Taiwan Topological and Nanostructured Materials," *All Partners Access Network*, undated.

Air Force Office of Scientific Research, *Funding Opportunity Announcement: FY20 Defense Established Program to Stimulate Competitive Research (DEPSCoR)*, undated.

Air Force Office of Scientific Research and the National Research Foundation of Korea, "AFOSR—NRF Korea Quantum Information Science and Technologies 2020–2022: Amplification of AFOSR BAA—FA9550-19-S-0003," undated.

ANR—*See* Agence Nationale de la Recherche.

ARO—*See* U.S. Army Research Office.

Australian Academy of the Humanities, Department of Industry and Science, *Measuring the Value of International Research Collaboration*, May 2015.

Baker Institute for Public Policy, Rice University, *Science Collaboration Across Borders*, Baker Institute Policy Report, No. 42, August 2009.

Berkman, Paul Arthur, "Could Science Diplomacy Be the Key to Stabilizing International Relations?" *Scientific American*, June 12, 2018.

Boeing, "Speeding Up Hypersonic Innovation with HIFiRE 4," webpage, August 10, 2017. As of February 1, 2021:
http://www.boeing.com/features/2017/08/hifire-08-17.page

Boekholt, Patries, Jakob Edler, Paul Cunningham, and Kieron Flanagan, *Drivers of International Collaboration in Research*, Technopolis Group and the University of Manchester, April 1, 2009.

Bozeman, Barry, and Jan Youtie, *The Strength in Numbers: The New Science of Team Science*, Princeton University Press, 2017.

BRO—*See* Basic Research Office.

Brown, Michael, and Pavneet Singh, *China's Technology Transfer Strategy: How Chinese Investments in Emerging Technology Enable a Strategic Competitor to Access the Crown Jewels of U.S. Innovation*, Defense Innovation Unit Experimental, January 2018.

Burt, Ronald S., "Structural Holes and Good Ideas," *American Journal of Sociology*, Vol. 110, No. 2, September 2004, pp. 349–399.

Burt, Ronald S., *Structural Holes: The Social Structure of Competition*, Harvard University Press, 1992.

Byerlee, Derek, and H. J. Dubin, "Crop Improvement in the CGIAR as a Global Success Story of Open Access and International Collaboration," *International Journal of the Commons*, Vol. 4, No. 1, 2010, pp. 452–480.

Capper, Peter, and Safa Kasap, eds., *Springer Handbook of Electronic and Photonic Materials*, Springer International Publishing, 2017.

Carter, Ashton B., "Fundamental Research: Memorandum for the Secretaries of the Military Departments," U.S. Department of Defense, May 24, 2010.

Casto, Matthew, "Trusted and Assured Microelectronics Program," DARPA ERI Summit, August 2020.

Centre National de la Recherche Scientifique [National Scientific Research Center], "International," webpage, undated. As of January 16, 2021:
http://www.cnrs.fr/en/international

CERN—*See* European Organization for Nuclear Research.

Chen, Yani, Yong Sun, Jiajun Peng, Junhui Tang, Kaibo Zheng, and Ziqi Liang, "2D Ruddlesden-Popper Perovskites for Optoelectronics," *Advanced Materials*, Vol. 30, No. 2, 2017.

Clery, Daniel, "New Review Slams Fusion Project's Management," *Science*, Vol. 343, No. 6174, February 28, 2014, pp. 957–958.

Clinton, William J., "Establishment of the National Science and Technology Council," Executive Order 12881, November 23, 1993.

CNRS—*See* Centre National de la Recherche Scientifique.

Code of Federal Regulations, Title 32, National Defense; Subtitle A, Department of Defense; Chapter I, Office of the Secretary of Defense; Subchapter M, Miscellaneous; Part 272, Administration and Support of Basic Research by the Department of Defense; Section 272.3, Definition of Basic Research.

Coleman, Mary Sue, "Letter to JCORE," Association of American Universities, January 27, 2020.

Consulate General of the People's Republic of China in New York, "National Basic Research Program of China (973 Program)," news release, March 5, 2016.

Cornell University, "International Engagement for Faculty," webpage, undated. As of July 29, 2021:
https://global.cornell.edu/operations/international-engagement-faculty

Council on Governmental Relations, *Framework for Review of Individual Global Engagements in Academic Research: Version 1.0*, January 14, 2020.

CREST Working Group, *Internationalisation of R&D—Facing the Challenge of Globalisation: Approaches to a Proactive International Policy in S&T*, November 27, 2007.

Culhane, J. L., and J. C. Worms, "Improving the Effectiveness of International Collaboration in Space Science," *Space Policy*, Vol. 17, 2001, pp. 179–186.

Curran, Kevin, "Is Taiwan a Ticking Time Bomb in the Semiconductor Supply Chain?" *The Diplomat*, February 27, 2021.

Danchev, Valentine, Andrey Rzhetsky, and James A. Evans, "Meta-Research: Centralized Scientific Communities Are Less Likely to Generate Replicable Results," *Elife*, Vol. 8, e43094, July 2, 2019.

Defence Science and Technology Group, Australian Government Department of Defence, "HIFIRE PROGRAM," webpage, undated. As of February 1, 2021:
https://www.dst.defence.gov.au/partnership/hifire-program

Defense Science Board, *Report of the Defense Science Board Task Force on Basic Research*, Office of the Under Secretary for Defense for Acquisition, Technology, and Logistics, January 2012.

Defense Science Board, *High Performance Microchip Supply*, Office of the Under Secretary of Defense for Acquisition, Technology, and Logistics, February 2005.

de Hoon, M. J. L., S. Imoto, J. Nolan, and S. Miyano, "Open Source Clustering Software," *Bioinformatics*, Vol. 20, No. 9, 2004, pp. 1453–1454. As of October 1, 2022:
http://bonsai.hgc.jp/~mdehoon/software/cluster/

DoD—*See* U.S. Department of Defense.

DoD High-Performance Computing Modernization Program, *HPCMP New Users' Guide: "Who May Run on HPCMP Resources?"* undated.

DOE—*See* U.S. Department of Energy.

DSB—*See* Defense Science Board.

Duffie, Warren Jr., "Arctic Ambition: ONR Joins International Mission to Monitor Conditions in Arctic Ocean," Office of Naval Research Public Affairs, September 14, 2017.

Edler, Jakob, John Rigby, and Barbara Jones, *Review of Human Frontier Science Program Final Report*, Human Frontiers Science Program, 2010.

ELI—*See* Extreme Light Infrastructure.

ERA Expert Group, *Opening to the World: International Cooperation in Science and Technology*, European Commission, EUR 23325 EN, 2008.

European Commission DG Communications Networks, Content & Technology, "Comparison of European and Non-European Regional Clusters in KETs [Key Enabling Technologies]: The Case of Semiconductors," webpage, 2013. As of August 1, 2021: https://ec.europa.eu/digital-single-market/en/news/comparison-european-and-non-european-regional-clusters-kets-case-semiconductors

European Organization for Nuclear Research, "Our Member States," webpage, undated. As of February 1, 2021: https://home.cern/about/who-we-are/our-governance/member-states

European Organization for Nuclear Research, "US Becomes Observer at CERN," press release, December 19, 1997. As of February 1, 2021: https://home.cern/news/press-release/cern/us-becomes-observer-cern

European Organization for Nuclear Research, "China," webpage, February 25, 2020. As of July 30, 2021: https://international-relations.web.cern.ch/stakeholder-relations/states/china

European Union, "Eurostat," webpage, undated. As of December 17, 2022: https://ec.europa.eu/eurostat/web/main/data/database

Extreme Light Infrastructure, "The ELI Project," webpage, undated. As of November 17, 2021: https://eli-laser.eu/the-eli-project/

Fedasiuk, Ryan, and Emily Weinstein, "Universities and the Chinese Defense Technology Workforce," *Center for Security and Emerging Technology*, December 2020.

Federal Accounting Standards Advisory Board, *FASAB Handbook of Federal Accounting Standards and Other Pronouncements, as Amended*, Version 19, June 30, 2020.

Federal Bureau of Investigation, "Foreign Government-Sponsored Talent Recruitment Plans, Such as China's Talent Plans, Incentivize Economic Espionage and Theft of Trade Secrets," PSA Number 20200716-2, July 16, 2020.

Federal Republic of Germany and the French Republic, Treaty Between the Federal Republic of Germany and the French Republic on Franco-German Cooperation and Integration, January 22, 2019.

Fermilab, "DUNE at LBNF: Countries and Institutions Participating in DUNE," webpage, undated. As of February 1, 2021: https://lbnf-dune.fnal.gov/about/countries-and-institutions-participating-in-dune/

Fermilab, "DUNE at LBNF: International Partnerships," webpage, undated. As of February 1, 2021: https://lbnf-dune.fnal.gov/about/international-partnerships/

Flagg, Melissa, and Paul Harris, "System Re-Engineering: A New Policy Framework for the American R&D System in a Changed World," Center for Security and Emerging Technology, September 2020.

Fleming, L., H. Greene, G. Li, M. Marx, and D. Yao, "Government-Funded Research Increasingly Fuels Innovation," *Science*, Vol. 364, No. 6446, June 21, 2019, pp. 1139–1141.

Freshwater, Dawn, Gwen Sherwood, and Vicki Drury, "International Research Collaboration: Issues, Benefits and Challenges of the Global Network," *Journal of Research in Nursing*, July 1, 2006.

Georghiou, Luke, "Global Cooperation in Research," *Research Policy*, Vol. 27, No. 6, September 1998, pp. 611–626.

Griffin, Michael D., *Actions for the Protection of Intellectual Property, Controlled Information, Key Personnel and Critical Technologies*, Memorandum for Under Secretary of Defense for Acquisition and Sustainment, Assistant Secretary of Defense for Acquisition, Service Acquisition Executives, Special Operations Command, March 20, 2019.

Griffin, Michael D., *Letter to Academic Colleagues*, U.S. Department of Defense, October 10, 2019.

Hesla, Leah, ed., "Fermilab, UNICAMP and São Paulo Research Foundation Collaborate on Major International Projects for Neutrino Research, Fermilab," webpage, March 13, 2020. As of February 1, 2021:
https://news.fnal.gov/2020/03/fermilab-unicamp-and-sao-paulo-research-foundation-collaborate-on-major-international-projects-for-neutrino-research/

HHS—*See* U.S. Department of Health and Human Services.

Hodiak, Justin, and Scott W. Harold, "Can China Become the World Leader in Semiconductors?" *The Diplomat*, September 25, 2020.

Holdren, John P, "Statement of the Honorable Dr. John P. Holdren, Director, Office of Science and Technology Policy, Executive Office of the President of the United States, to the Subcommittee on Oversight and Investigations, House Committee on Foreign Affairs," November 2, 2011.

Hsin-fang, Lee, and William Hetherington, "Taiwan, US to Focus on Semiconductors," *Taipei Times*, November 22, 2020, p. 1.

JASON, *Fundamental Research Security*, MITRE Corporation, December 2019.

Jia, Hepeng, "China's Science Ministry Gets Power to Attract More Foreign Scientists: Science Super-Ministry Gains New Remit to Develop Policies That Benefit International Researchers," *Nature Index*, March 23, 2018.

Jia, Hepeng, "China Reshuffles Science Governance: Moves May Generate Conflict Between Basic and Applied Research Programs," *C&EN*, Vol. 96, No. 18, April 25, 2018.

Joint Committee on the Research Environment, National Science and Technology Council, Executive Office of the President, *Summary of the 2019 White House Summit of the Joint Committee on the Research Environment (JCORE)*, November 2019.

Joske, Alex, "The China Defence Universities Tracker," Australian Strategic Policy Institute, November 2019. As of August 3, 2021:
https://www.aspi.org.au/report/china-defence-universities-tracker

Joske, Alex, *Hunting the Phoenix: The Chinese Communist Party's Global Search for Technology and Talent*, Australian Strategic Policy Institute: International Cyber Policy Centre, No. 35, August 2020.

Kania, John, and Mark Kramer, "Collective Impact," *Stanford Social Innovation Review*, Winter 2011.

Katz, J. Sylvan, and Ben Martin, "What Is Research Collaboration?" *Research Policy*, Vol. 26, No. 1, 1997, pp. 1–18.

Kerr, Sari Pekkala, and William R. Kerr, "Global Collaborative Patents," *The Economic Journal*, Vol. 128, No. 612, 2018, pp. F235–F272.

Khan, Saif M., and Alexander Mann, "AI Chips: What They Are and Why They Matter," Center for Security and Emerging Technology, April 2020.

Knight, Will, "Intel Wants to Revive U.S. Chipmaking—But It Has to Catch Up First," *Wired*, March 24, 2021.

Krasnyak, Olga, "How U.S.-Soviet Scientific and Technical Exchanges Helped End the Cold War," *American Diplomacy*, November 2019.

Kuhn, Thomas S., *The Structure of Scientific Revolutions*, University of Chicago Press, 1962.

Lami, Stefano, "Challenges and New Requirements for International Collaborations," *Science & Diplomacy*, Vol. 6, No. 2, June 2017.

Laserlab-Europe, "Using Our Facilities," webpage, undated. As of February 1, 2021:
https://www.laserlab-europe.eu/transnational-access

Lee, Jenny J., and John P. Haupt, "Winners and Losers in US-China Scientific Research Collaboration," *Higher Education*, Vol. 80, July 6, 2020, pp. 57–74.

Lee, Sooho, and Barry Bozeman, "The Impact of Research Collaboration on Scientific Productivity," *Social Studies of Science*, Vol. 35, No. 5, October 1, 2005, pp. 673–702.

Lungeanu, Alina, and Noshir S. Contractor, "The Effects of Diversity and Network Ties on Innovations: The Emergence of a New Scientific Field," *American Behavioral Scientist*, Vol. 59, No. 5, May 1, 2015, pp. 548–564.

McCants, Carl E., "IARPA Trusted Integrated Chips (TIC) Program," Office of the Director of National Intelligence, August 17, 2016. As of June 2, 2021:
https://www.ndia.org/-/media/sites/ndia/meetings-and-events/divisions/systems-engineering/past-events/trusted-micro/2016-august/mccants-carl.ashx

Mattis, Jim, *Summary of the 2018 National Defense Strategy of the United States of America: Sharpening the American Military's Competitive Edge*, U.S. Department of Defense, undated.

MITRE, "USD(R&E) Top 10 Technology Focus Areas," webpage, undated. As of September 21, 2020:
https://aida.mitre.org/top-10-technology-areas/

Mölling, Christian, *Defense Innovation and the Future of Transatlantic Strategic Superiority: A German Perspective*, German Marshall Fund of the United States, March 23, 2018.

Muldavin, Jeremy, "DoD Trusted and Assured Microelectronics Summary," NDIA Electronics Division Meeting, February 7, 2019.

National Academy of Engineering, *Scientific Communication and National Security*, National Academies Press, 1982.

National Academy of Sciences, National Academy of Engineering, and Institute of Medicine, "Research Associateship Programs: Participating Agencies," webpage, undated. As of February 1, 2021:
http://nrc58.nas.edu/RAPLab10/Opportunity/Programs.aspx

National Academy of Sciences, National Academy of Engineering, and Institute of Medicine, *Examining Core Elements of International Research Collaboration: Summary of a Workshop*, National Academies Press, 2011.

National Bureau of Statistics, People's Republic of China, "Science and Technology Development by Leaps and Innovation Leading a New Chapter—Seventh Series of Reports on Economic and Social Development Achievements in the 70th Anniversary of the Founding of New China [科技发展大跨越 创新引领谱新篇--新中国成立70周年经济社会发展成就系列报告之七]," 2019. As of August 4, 2021:
http://www.stats.gov.cn/tjsj/zxfb/201907/t20190723_1680979.html

National Center for Science and Engineering Statistics, *Survey of Federal Funds for Research and Development, Fiscal Years 2018–19*, January 2020.

National Defense Authorization Act for Fiscal Year 2021, Section 276, Microelectronics and National Security.

National Defense Authorization Act for Fiscal Year 2021, Sections 9901–9908, Creating Helpful Incentives to Produce Semiconductors (CHIPS) for America Act.

National Nanotechnology Initiative, "Highlights from the Nanoelectronics for 2020 and Beyond (Nanoelectronics) NSI [Nanotechnology Signature Initiative]," April 2017.

National Natural Science Foundation of China, *2018 Annual Report*, English version, undated.

National Natural Science Foundation of China, "The International Workshop on Science Funding Strategies and Policies Successfully Held," webpage, undated.
https://www.nsfc.gov.cn/english/site_1/news/A1/2020/04-27/118.html

National Natural Science Foundation of China, "NSFC at a Glance," webpage, undated. As of February 2, 2021:
http://www.nsfc.gov.cn/english/site_1/about/6.html

National Natural Science Foundation of China, *National Natural Science Fund Guide to Programs 2019*, English version, undated.

National Natural Science Foundation of China, "NSFC Secretary-General Han Yu Attends the 94th Meeting of IIASA Council and Visits Norway," webpage, undated. As of February 2, 2021:
http://www.nsfc.gov.cn/english/site_1/news/A1/2020/04-27/180.html

National Natural Science Foundation of China, "NSFC Vice President Wang Chengwen Attends the 6th World Conference on Research Integrity," webpage, undated. As of February 2, 2021:
http://www.nsfc.gov.cn/english/site_1/news/A1/2020/04-27/179.html

National Natural Science Foundation of China, "NSFC Vice President Wang Chengwen Attends the 16th meeting of China-Japan Committee on Intergovernmental Science and Technology Cooperation and Visits JSPS," webpage, undated. As of February 2, 2021:
http://www.nsfc.gov.cn/english/site_1/news/A1/2018/11-12/122.html

National Natural Science Foundation of China, *International Cooperation Bureau, List of NSFC-Signed Cooperation Agreements (MOUs)* [NSFC签署合作协议(谅解备忘录)情况一览表], webpage, March 3, 2020. As of February 2, 2021:
http://bic.nsfc.gov.cn/Show.aspx?AI=1261

National Oceanic and Atmospheric Administration, Office of Space Commerce, "Revised Satellite Export Control Rules Published," webpage, May 13, 2014. As of July 27, 2021:
https://www.space.commerce.gov/revised-satellite-export-control-rules-published/

National Research Council, *Funding a Revolution: Government Support for Computing Research*, National Academies Press, 1999.

National Research Council, *Materials Research to Meet 21st-Century Defense Needs*, National Academies Press, 2003.

National Research Council, *Assessment of Department of Defense Basic Research*, National Academies Press, 2005.

National Research Council, *Space Science and the International Traffic in Arms Regulations: Summary of a Workshop*, National Academies Press, 2008.

National Research Council, *Beyond "Fortress America": National Security Controls on Science and Technology in a Globalized World*, National Academies Press, 2009.

National Research Council, *S&T Strategies of Six Countries: Implications for the United States*, National Academies Press, 2010.

National Research Council, *Optics and Photonics: Essential Technologies for Our Nation*, National Academies Press, 2013.

National Research Council, *Strategic Engagement in Global S&T: Opportunities for Defense Research*, National Academies Press, 2014.

National Science Board, National Science Foundation, *Research and Development: Essential Foundation for U.S. Competitiveness in a Global Economy—A Companion to Science and Engineering Indicators—2008*, NSB-08-3, 2008.

National Science Board, National Science Foundation, *Publications Output: U.S. Trends and International Comparisons*, NSB-2020-6, December 17, 2019.

National Science Board, National Science Foundation, *Research and Development: U.S. Trends and International Comparisons*, NSB-2020-3, January 15, 2020.

National Science Foundation, "About the National Science Foundation," webpage, undated. As of February 2, 2021:
https://www.nsf.gov/about/

National Science Foundation, *Accelerating Research Through International Network-to-Network Collaborations (AccelNet): Program Solicitation NSF 21-511*, undated.

National Science Foundation, *Centers for Chemical Innovation: Phase I Awards and New/Renewal Phase II Centers: Program Solicitation NSF 20-574*, undated.

National Science Foundation, "Funding: Division of Environmental Biology—Dimensions of Biodiversity," webpage, undated. As of February 2, 2021:
https://www.nsf.gov/funding/pgm_summ.jsp?pims_id=503446

National Science Foundation, "International Gemini Observatory," webpage, undated. As of February 1, 2021:
https://www.nsf.gov/funding/pgm_summ.jsp?pims_id=5651&org=MPS&from=home

National Science Foundation, *NSF FY 2021 Budget Request to Congress: High Luminosity-Large Hadron Collider Upgrade (HL-LHC)*, undated.

National Science Foundation, "Science and Engineering Indicators," webpage, undated. As of July 20, 2021:
https://www.nsf.gov/statistics/seind/

National Science Foundation, "Science and Technology Centers: Integrative Partnerships: Program Solicitation NSF 19-567," webpage, undated. As of February 2, 2021:
https://www.nsf.gov/pubs/2019/nsf19567/nsf19567.htm

National Science Foundation, "Who We Are," webpage, undated. As of February 2, 2021:
https://www.nsf.gov/about/who.jsp

National Science Foundation, "The Excitement Behind a Renewed Agreement," webpage, May 7, 2015. As of February 2, 2021:
https://www.nsf.gov/mps/ad_blogs/
the_excitement_behind_a_renewed_agreement_may2015.jsp

National Science Foundation, *Building the Future: Investing in Discovery and Innovation: NSF Strategic Plan for Fiscal Years (FY) 2018–2022*, NSF 18-45, February 2018.

National Science Foundation, "NSF Announces Changes to Overseas Offices," press release 18-003, February 21, 2018.

National Science Foundation, "Master Government List of Federally Funded R&D Centers," webpage, June 2020. As of February 2, 2021:
https://www.nsf.gov/statistics/ffrdclist/

National Science Foundation, *Proposal and Award Policies and Procedures Guide*, effective June 1, 2020. As of February 2, 2021:
https://www.nsf.gov/publications/pub_summ.jsp?ods_key=pappg

National Science Foundation, National Radio Astronomy Observatory, "The NRAO Enables Forefront Research into the Invisible Universe," webpage, undated. As of February 2, 2021:
https://public.nrao.edu/about/

National Science Foundation, Office of International Science and Engineering, "Counterpart Science Funding Agencies," webpage, undated. As of February 1, 2021:
https://www.nsf.gov/od/oise/counterpart-funding-orgs.jsp

National Science Foundation, Office of International Science and Engineering, "International Activities at NSF," webpage, undated. As of February 2, 2021:
https://www.nsf.gov/od/oise/IntlCollaborations/index.jsp

National Science Foundation, Office of International Science and Engineering, "International Research Experiences for Students (IRES)," webpage, undated. As of February 2, 2021:
https://www.nsf.gov/funding/pgm_summ.jsp?pims_id=505656&org=OISE&from=home

National Science Foundation, Office of International Science and Engineering, "NSF MULTIPLIER—MULTIPlying Impact Leveraging International Expertise in Research Missions," webpage, undated. As of February 1, 2021:
https://www.nsf.gov/od/oise/multiplier.jsp

National Science Foundation, Office of International Science and Engineering, "Partnerships for International Research and Education (PIRE)," webpage, undated. As of July 30, 2021:
https://www.nsf.gov/funding/pgm_summ.jsp?pims_id=505038&org=OISE&from=home

National Science Foundation, Office of the Director, *Personnel Policy on Foreign Government Talent Recruitment Programs*, July 11, 2019.

Nokkala, Terhi, *Internal Collaboration Rules in International R&D Collaboration Projects—Analysis of Seven NEST Projects*, NEMO, NEMO Working Paper #15, January 2009.

Normile, Dennis, "China Asserts a Firm Grip on Research Data," *Science*, April 9, 2018.

NRC—*See* National Research Council.

NSB—*See* National Science Board.

NSF—*See* National Science Foundation.

NSFC—*See* National Natural Science Foundation of China.

Nuclear Science User Facilities, "About Us," webpage, undated. As of February 2, 2021:
https://nsuf.inl.gov/Page/about

Nuclear Science User Facilities, "Rapid Turnaround Experiments," webpage, undated. As of February 2, 2021:
https://nsuf.inl.gov/Page/rte

OECD—*See* Organisation for Economic Co-operation and Development.

OECD.Stat, homepage, undated. As of December 17, 2022:
https://stats.oecd.org/

Office of Naval Research, "ONR Global Funding Opportunities," webpage, undated. As of February 1, 2021:
https://www.onr.navy.mil/Science-Technology/ONR-Global/funding-opportunities

Office of Naval Research, Army Research Office, and Air Force Office of Scientific Research, "Fiscal Year (FY) 2020 Department of Defense Multidisciplinary Research Program of the University Research Initiative," undated.

Office of Science and Technology Policy, "Enhancing the Security and Integrity of America's Research Enterprise," briefing, July 2020.

Office of the Director for International Cooperation, Office of the Under Secretary of Defense for Acquisition, Technology and Logistics, *International Cooperation in Acquisition, Technology and Logistics (IC in AT&L) Handbook*, 7th ed., May 2012.

Office of the State Council, People's Republic of China, "Notice of the General Office of the State Council on Issuing the Measures for the Administration of Scientific Data [国务院办公厅关于印发科学数据管理办法的通知]," April 2, 2018. As of August 23, 2022:
http://www.gov.cn/zhengce/content/2018-04/02/content_5279272.htm

O'Keeffe, Kate, and Melissa Korn, "China's Military Sends More Scholars Abroad, at Times Without Schools' Knowledge," *Wall Street Journal*, October 29, 2018.

ONR—*See* Office of Naval Research.

Organisation for Economic Co-operation and Development, "Gross Domestic Expenditure on R&D by Sector of Performance and Type of R&D," *OECD.Stat*, undated. As of November 15, 2021:
https://stats.oecd.org/Index.aspx?DataSetCode=MSTI_PUB#

OSTP—*See* Office of Science and Technology Policy.

Owens, Brian, "The Benefits and Challenges of International Research Collaboration," *University Affairs*, November 7, 2018.

Pain, Elisabeth, "Tips for Successful International Collaborations," *Science*, February 15, 2013.

Payette, Julie, "Research and Diplomacy 350 Kilometers Above the Earth," *Science & Diplomacy*, December 10, 2012.

Phillips, Robert F., *Administrative History of the European Office of Aerospace Research and Development, 1952–1975*, Air Force Systems Command, 1977.

Platzer, Michaela D., John F. Sargent, Jr., and Karen M. Sutter, *Semiconductors: U.S. Industry, Global Competition, and Federal Policy*, Congressional Research Service, R46581, 2020.

Powell, Walter W., Kenneth K. Koput, and Laurel Smith-Doerr, "Interorganizational Collaboration and the Locus of Innovation: Networks of Learning in Biotechnology," *Administrative Science Quarterly*, Vol. 41, No. 1, 1996, pp. 116–145.

Primeri, Emilia, Emanuela Reale, Benedetto Lepori, Philippe Laredo, Maria Nedeva, and Duncan Thomas, "Measuring the Opening of National R&D Programs: What Indicators for What Purposes?" *Research Evaluation*, Vol. 23, No. 4, October 2014, pp. 312–326.

Princeton Plasma Physics Laboratory, "U.S.-China Collaboration Makes Excellent Start in Optimizing Lithium to Control Fusion Plasmas," press release, June 22, 2017.

Public Law 94-282, National Science and Technology Policy, Organization, and Priorities Act of 1976.

Reagan, Ronald, *National Policy on the Transfer of Scientific, Technical, and Engineering Information*, National Security Decision Directive 189, September 21, 1985.

Riesselmann, Kurt, "UK Invests £65 Million in International Science Projects Hosted by Fermilab," Fermilab, press release, January 22, 2020.

Royal Society and American Society for the Advancement of Science, "New Frontiers in Science Diplomacy: Navigating the Changing Balance of Power," January 2010.

Rulison, Larry, "NY: Demise of G450C Wasn't over Money," *Times Union*, January 13, 2017.

Saha, Chandra Nath, and Sanjib Bhattacharya, "Intellectual Property Rights: An Overview and Implications in Pharmaceutical Industry," *Journal of Advanced Pharmaceutical Technology & Research*, Vol. 2, No. 2, 2011, pp. 88–93.

Sanger, David E., Catie Edmondson, David McCabe, and Thomas Kaplan, "Senate Poised to Pass Huge Industrial Policy Bill to Counter China," *New York Times*, June 7, 2021.

Sargent, John F., Jr., *Global Research and Development Expenditures: Fact Sheet*, Congressional Research Service, June 27, 2018.

Schiermeier, Quirin, "Horizon 2020 by the Numbers: How €60 Billion Was Divided up Among Europe's Scientists," *Nature*, December 22, 2020.

Semiconductor Industry Association, *2020: State of the U.S. Semiconductor Industry*, undated.

SESAME—*See* Synchrotron-light for Experimental Science and Applications in the Middle East.

Shead, Sam, "The Global Chip Shortage Is Starting to Have Major Real-World Consequences," *CNBC*, May 7, 2021.

Silicon Europe, "Silicon Europe: Cluster Alliance for European Micro- and Nanoelectronics Industry," press release, October 8, 2012.

Stamm, Aurelia, and Andreas Figueroa, "Effective International Science, Technology and Innovation Collaboration: From Lessons Learned to Policy Change," in OECD, *Meeting Global Challenges Through Better Governance: International Co-operation in Science, Technology and Innovation*, OECD Publishing, 2012.

Stracqualursi, Veronica, and Sheena Jones, "Harvard Professor Among Three Charged with Lying About Chinese Government Ties," *CNN*, January 28, 2020.

Synchrotron-light for Experimental Science and Applications in the Middle East, "What Is SESAME?" webpage, undated. As of February 2, 2021:
https://www.sesame.org.jo/about-us/what-is-sesame

Tanner, Murray Scot, "Beijing's New National Intelligence Law: From Defense to Offense," *Lawfare*, July 20, 2017.

Thompson, Haydn, Emilio Lora-Tamayo, Damm Werner, Jean-Luc Dormoy, Leonard Hobbs, Margriet Jansz, Tomasz Kosmider, and Wolfgang Pribyl, "Final Evaluation of the ARTEMIS and ENIAC Joint Undertaking (2008-2013) Operating Under FP7, Directorate-General for Communications Networks, Content and Technology (European Commission)," June 2017.

Trump, Donald J., Presidential Memorandum on United States Government-Supported Research and Development National Security Policy, NSPM-33, January 14, 2021.

United Nations Conference on Trade and Development, "China—Introduces 'Measures for the Overseas Transfers of Intellectual Property Rights (Trial),'" webpage, March 29, 2018. As of July 30, 2021:
https://investmentpolicy.unctad.org/investment-policy-monitor/measures/3252/china-introduces-measures-for-the-overseas-transfers-of-intellectual-property-rights-trial-

United Nations Environment Program, International Ecosystem Management Partnership, "UN Environment-NSFC Bilateral Cooperation Meeting," January 29, 2017.

United States of America Cyberspace Solarium Commission, "Building a Trusted ICT Supply Chain: CSC White Paper #4," October 2020.

U.S. Air Force Research Laboratory, Air Force Office of Scientific Research, *AFOSR International Office FY19 Annual Report*, undated.

U.S. Air Force Research Laboratory, "#BasicResearch Chatter—An Opportunity to Chat About the AFRL Engineer and Scientist Exchange Program (ESEP)," *Afrl.dod.live.mil*, December 14, 2020. As of February 1, 2021:
https://afrl.dodlive.mil/category/researchers/

U.S. Army, Army Research Laboratory, "Partnering Mechanisms | Open Campus," webpage, undated. As of July 30, 2021:
https://www.arl.army.mil/opencampus/partnering

U.S. Army Research Office, *2019 ARO in Review*, U.S. Army Combat Capabilities Development Command, Army Research Laboratory, undated.

U.S.-China Economic and Security Review Commission, *2019 Report to Congress of the U.S.-China Economic and Security Review Commission*, U.S. Government Publishing Office, November 2019.

U.S. Code, Title 10, Section 301, Definitions, January 3, 2012.

U.S. Code, Title 10, Section 2350a, Cooperative Research and Development Agreements, January 7, 2011.

U.S. Code, Title 10, Section 2358, Research and Development Projects, January 5, 2015.

U.S. Code, Title 22, Section 2767, Arms Export Control Act: Authority of President to Enter into Cooperative Projects with Friendly Foreign Countries, 2012.

U.S. Code, Title 42, Section 1861-75, National Science Foundation Act of 1950, 1950.

U.S. Department of Defense, "DAU (Defense Acquisition University) Glossary," webpage, undated. As of December 12, 2021:
https://www.dau.edu/glossary/SiteAssets/download.aspx

U.S. Department of Defense, *International S&T Engagement Strategy*, July 2014.

U.S. Department of Defense, "DoD Announces BARI Award for US-UK Collaboration on Human-Machine Teaming," press release, September 25, 2018.

U.S. Department of Defense, DoD Instruction 5530.03, *International Agreements*, December 4, 2019.

U.S. Department of Defense, *Department of Defense Directive 5137.02: Under Secretary of Defense for Research and Engineering (USD[R&E])*, July 15, 2020.

U.S. Department of Defense, *Department of Defense International Science and Technology Engagement Strategy: A Unified Approach to Strengthen Alliances and Attract New Partners*, October 2020.

U.S. Department of Defense, Basic Research Office, "Our Mission," webpage, undated. As of February 1, 2021:
https://basicresearch.defense.gov/About/

U.S. Department of Defense, Washington Headquarters Services, Acquisition Directorate, *Funding Opportunity Announcement #WHS-AD-FOA-18: Minerva Research Initiative*, undated.

U.S. Department of Energy, "Definition," webpage, undated. As of February 2, 2021:
https://science.osti.gov/User-Facilities/Policies-and-Processes/Definition

U.S. Department of Energy, *Bilateral DOE Laboratory Utilization Agreement*, May 26, 2010.

U.S. Department of Energy, *Fiscal Year 2017 Annual Performance Report; Fiscal Year 2019 Annual Performance Plan*, undated.

U.S. Department of Energy, *Foreign Engagements with DOE National Laboratories*, DOE P 485.1A, December 13, 2019.

U.S. Department of Energy, *Foreign Government Sponsored or Affiliated Activities*, DOE O 486.1A, September 4, 2020.

U.S. Department of Energy, National Energy Technology Laboratory, "NETL Plays Key Role in DOE's New Lab Partnering Service Web Experience," webpage, July 11, 2018. As of February 2, 2021:
https://www.netl.doe.gov/node/5451

U.S. Department of Energy, Office of Energy Efficiency & Renewable Energy, "How to Partner with the National Labs," webpage, undated. As of July 30, 2021:
https://www.energy.gov/eere/wind/how-partner-national-labs

U.S. Department of Energy, Office of Fossil Energy and Carbon Management, "Bilateral Agreements with China," webpage, undated. As of July 30, 2021:
https://www.energy.gov/fe/services/international-cooperation/bilateral-agreements-china

U.S. Department of Energy, Office of International Affairs, "International Affairs Initiatives," webpage, undated. As of February 2, 2021:
https://www.energy.gov/ia/international-affairs-initiatives

U.S. Department of Energy, Office of International Affairs, "Welcome to DOE International," webpage, April 18, 2014. As of February 2, 2021:
https://www.energy.gov/ia/articles/welcome-doe-international

U.S. Department of Energy, Office of Nuclear Energy, "Bilateral Cooperation," webpage, undated. As of February 2, 2021:
https://www.energy.gov/ne/nuclear-reactor-technologies/international-nuclear-energy-policy-and-cooperation/bilateral

U.S. Department of Energy, Office of Nuclear Energy, "International Nuclear Energy Policy and Cooperation," webpage, undated. As of February 2, 2021:
https://www.energy.gov/ne/nuclear-reactor-technologies/international-nuclear-energy-policy-and-cooperation

U.S. Department of Energy, Office of Science, "About the Office of Science," webpage, undated. As of February 2, 2021:
https://www.energy.gov/science/about-office-science

U.S. Department of Energy, Office of Science, *Basic Research Needs for Microelectronics: Report of the Office of Science Workshop on Basic Research Needs for Microelectronics, October 3–25, 2018*, undated.

U.S. Department of Energy, Office of Science, *FY 2020 Continuation of Solicitation for the Office of Science Financial Assistance Program, Funding Opportunity Announcement (FOA) Number: DE-FOA-0002181*, November 1, 2019.

U.S. Department of Health and Human Services, Office of Research Integrity, "Need for Collaboration," webpage, undated. As of July 30, 2021:
https://ori.hhs.gov/education/products/niu_collabresearch/collabresearch/need/need.html

U.S. Department of State, "Military-Civil Fusion and the People's Republic of China," fact sheet, May 28, 2020.

U.S. General Accounting Office, "Lessons Learned from SEMATECH," GAO/RCED-92-283, September 1992.

U.S. Government Accountability Office, *Federal Research: Agencies Need to Enhance Policies to Address Foreign Influence*, GAO-21-130, December 2020.

U.S. House Committee on Appropriations, *Report to Accompany H.R. 9221 (Department of Defense Appropriation Bill, 1966)*, Report No. 528, June 19, 1965.

U.S. Senate Permanent Subcommittee on Investigations, "Threats to the U.S. Research Enterprise: China's Talent Recruitment Plans," staff report, 2019. As of July 31, 2021:
https://www.hsgac.senate.gov/subcommittees/investigations/hearings/securing-the-us-research-enterprise-from-chinas-talent-recruitment-plans

Varas, Antonio, Raj Varadarajan, Jimmy Goodrich, and Falan Yinug, "Government Incentives and U.S. Competitiveness in Semiconductor Manufacturing," *BCG and Semiconductor Industry Association*, September 2020.

Wagner, Caroline S., "International Collaboration in Science and Technology: Promises and Pitfalls," in Louk Box and Rutger Engelhard, eds., *Science and Technology Policy for Development, Dialogues at the Interface*, Anthem Press, 2006.

Wagner, Caroline S., Irene T. Brahmakulam, Brian A. Jackson, Anny Wong, and Tatsuro Yoda, *Science & Technology Collaboration: Building Capacity in Developing Countries?* RAND Corporation, MR-1357.0-WB, 2001. As of July 28, 2021:
https://www.rand.org/pubs/monograph_reports/MR1357z0.html

Watson, James D., *The Double Helix: A Personal Account of the Discovery of the Structure of DNA*, Touchstone, 2001.

Weinbaum, Cortney, Eric Landree, Marjory S. Blumenthal, Tepring Piquado, and Carlos Ignacio Gutierrez Gaviria, *Ethics in Scientific Research: An Examination of Ethical Principles and Emerging Topics*, RAND Corporation, RR-2912-IARPA, 2019. As of July 27, 2021:
https://www.rand.org/pubs/research_reports/RR2912.html

White House, "Fact Sheet: Securing America's Critical Supply Chains," February 24, 2021.

Working Group on Science and Technology in U.S.-China Relations, *Meeting the China Challenge: A New American Strategy for Technology Competition*, UC San Diego School of Global Policy and Strategy, November 16, 2020.

World Nuclear News, "DOE Announces Funds for Overseas Fusion Research," webpage, March 11, 2019. As of February 2, 2021:
https://www.world-nuclear-news.org/Articles/DOE-announces-funds-for-overseas-fusion-research

Wu, Debby, "Engineers Found Guilty of Stealing Micron Secrets for China," *Bloomberg*, June 12, 2020.

Yang, Fangxu, Shanshan Cheng, Xiaotao Zhang, Xiaochen Ren, Rongjin Li, Huanli Dong, and Wenping Hu, "2D Organic Materials for Optoelectronic Applications," *Advanced Materials*, Vol. 30, No. 2, 2018.